The Green Book of Household Hints

The Green Book of

Household Hints

Keeping an Efficient and Ecologically Sound Home

Marjorie Harris

Firefly Books

Published by Firefly Books (U.S.) Inc. 2001

First Printing

U.S. CATALOGING IN PUBLICATION DATA
(Library of Congress Standards)

Harris, Marjorie.
The green book of household hints : keeping an efficient and ecologically sound home / Marjorie Harris. –1st ed.
[224] p. : ill. ; cm.
Includes index.
Summary : Useful and environmentally responsible tips for housekeeping.
ISBN 1-55209-600-9
1.House cleaning. 2.Recycling (Waste, etc.). 3.Cleaning. I.Title.
648.5 21 2001

Published in the United States in 2001 by
Firefly Books (U.S.) Inc.
P.O. Box 1338, Ellicott Station
Buffalo, New York, USA
14205

A Word of Caution
None of the suggestions offered in this book is intended to take the place of advice available from qualified professionals. In particular, readers who find themselves confronting a medical emergency should obtain the help of a qualified medical practitioner at the earliest possible opportunity. The advice given with respect to household repairs is valid to the best of the author's knowledge, but obviously none of the topics is dealt with in sufficient depth to cover all possible eventualities. If the advice is not working, the reader should seek the assistance of an expert.

Parts of this book have previously appeared in Better House and Planet, *published by Key Porter Books Limited in 1991.*

Published in Canada in 2001 by Key Porter Books Limited.

Design: Peter Maher
Electronic formatting: Heidi Palfrey

Printed and bound in Canada

PREFACE

I became an environmentalist when I first read Rachel Carson's *Silent Spring* in the early 1960s. As a neophyte journalist and a mother with young children, it became important not only to write about the environment, but to try and live an ecologically correct life. I did stories about our ventures into eating health foods, what was happening to our city's water supply and the effect of pollution on our community. Eventually my editor said, "Do you have to write these depressing stories? Stop." Occasionally the magazine I worked for pronounced the death of a Great Lake or reported that birds were dying mysteriously, but the populace didn't seem to want to take up arms in outrage, only the usual lunatic fringe.

I counted myself among that lunatic fringe. We didn't drink tap water, my kids didn't eat stuff advertised on television, we bought groceries from the health food store and more recently from a food co-op. We lived very much as my parents lived during the Depression and the Second World War. We all survived it quite nicely. We learned to reuse, recycle and reduce the amount of waste we produced—much like the battle cry of contemporary environmentalists.

Today it's still only relatively easy to run an ecologically correct house. Most municipalities have a toxic waste program—they'll pick up your toxic wastes and get rid of them as safely as possible. (If you don't have one in your town, lobby to get one started.) There are dozens, soon to be hundreds, of so-called "Green" products on the market, which means you will have to be even more vigilant than before about reading the list of contents. One of the great ironies, for instance, is the environmentally friendly can of hair spray. It's still almost indestructible garbage even if it doesn't contain CFCs (chlorofluorocarbons), which destroy the ozone layer.

Some people find it hard to get along without plastic bags. I still have one or two around the house to carry things in. Our garbage output is so small that they aren't very useful as garbage bags. We compost, buy in bulk and avoid packaging as much as possible. And we reuse most things except junk mail (that goes back to the addressee with "Junk Mail" written on it). But, and there's always a but, there are times when a plastic bag or container comes in handy.

When a cleaning job is heavy-duty enough to require something as strong as household ammonia or even bleach, use the alternatives first: baking soda, salt, vinegar or borax. Borax is marvelous and is suited for endless jobs. I've pointed out which substances are dangerous. Use them with care: Wear rubber gloves when you are handling them and do so only in a well-ventilated space. And use them as little as possible.

Any references to detergent are to nonphosphate detergents; references to soap are to pure soaps in flake, powdered or liquid form. Any reference to vinegar, unless specified, is the cheapest white vinegar you can buy. Occasionally you'll find that paper towels are essential for a particular job. At least try to find recycled paper towels. Alum is mentioned in several hints. It isn't bad in itself, but sometimes the methods of extracting it are dangerous to the environment. In this case you have to weigh that against convenience.

You can help by avoiding chemicals. This book is an attempt to put together all the useful information available on how to run a house efficiently, economically and—what's most important—ecologically. Look for the green house in the margin. The information was gathered from friends, neighbors, books new and old, newspapers, magazines and experience.

To use the book, look up any topic alphabetically. If you can't find it, check the index and all the other subjects closely connected to the one you're interested in. For instance, if you want to look up food tips, look under specific categories such as Corn on the Cob or Olive Oil. If that fails, check out the index: anything to do with food or cooking tips will be listed under the general heading, Food.

On our street we started a neighborhood ecology group called Grassroots Albany. We meet every month and work out how we can make our environment a little safer. We've taken on garbage and recycling projects, lobbied with letter-writing campaigns and planted trees. We have a letterhead that we pass out along with a kit of sample letters, projects and information on how to get started. If you are interested in starting your own group, write to Grassroots Albany, 211 Albany Avenue, Toronto M5R 3C7, Canada.

I am grateful for dedicated and imaginative research on the part of Loral Dean, Karen O'Reilly, Sean O'Malley, Peggy Beatty, Juliet Mannock and Marge Stibbards.

Marjorie Harris, 1991

ADHESIVES (See Glue and Gluing.)

AIR CONDITIONERS

Air conditioners are a luxury in a world of rapidly rising air pollution. They contain CFCs to absorb heat from inside and transfer it. If there's a leak, there's a problem. When CFCs are evaporated into the atmosphere they contribute to ozone depletion.

Air conditioners also use a lot of energy. You might consider the following alternatives before you install one:

- Keep shades or curtains drawn, and doors and windows closed during the day to keep in cool air—opening doors and windows actually lets heat into the house. Open up the house for ventilation only in the late evening or very early morning.
- Turn on fans to circulate air.
- Plant rapid-growing shade trees around your house. By drawing moisture from deep in the soil and shedding it through their leaves, they cool the atmosphere around them and also provide welcome shade. The trees should have deep roots (maples are shallow-rooted trees); deciduous trees will allow sun into your home in the winter.
- Foundation plantings will keep your basement cool and this in turn will lower the temperature of your house slightly. Make sure they are 2 ft (60 cm) from the foundation walls, though.
- Consider buying a heat pump, which transfers hot air from inside your house and disperses it outdoors in summer.

If you absolutely must have an air conditioner, be sure to fit the right one to your needs. If you have a small home or modest cooling requirements, window units may be the best option. For a larger home, central air will be more efficient. A qualified air conditioner technician will be able to advise you on your needs. Be sure to check the energy efficiency of the air conditioner. High-efficiency models may cost a little more but will save you money in the long run.

Here is some helpful advice for window units:

- Match the capacity of the unit to the size of the room. A unit that's too small will run too long; one that's too large will cool the air so rapidly that the moisture vapor won't be removed, resulting in a buildup of humidity.
- Run the unit constantly; turning it off and on takes much more energy.
- Avoid putting the unit in an area where the sun beats down on it.
- Before you clean the unit, unplug it. Vacuum the front on a regular basis. Once a year, remove the cover and vacuum behind it. Do this before you reinstall it each year.
- Test it for leaks by running it on a warm, humid day. The unit will create some water, but it should drain to the back where it will evaporate except on the most humid days of summer. The front coil should be damp from top to bottom without dripping. On a very humid day, it will drip. Never use a drill on the unit even to rechannel water. Get it done professionally.
- If the unit isn't cooling properly, the filter may need either cleaning or replacing. It's located behind the front panel on the inside of the air conditioner. Loosen the panel and remove the filter. The easiest way to clean it is by vacuuming or washing in a mild detergent and warm water.
- If the unit drips into the room, check to make sure it's draining outside properly.
- A musty, stale odor means that the drain hole under the evaporator area is clogged. Remove the front panel, then clean it with a bit of wire from a coat hanger or use a long, narrow wire bottlebrush.
- Tape a thermometer to the air supply vent. Then do the same to the return air vent. There should be a difference of 17° to 20° on the Fahrenheit scale (6° to 8° on the Celsius scale).
- In the winter time, remove the units or cover and seal them.

For central air:

- Central air conditioners should be inspected, cleaned and tuned by a professional every two or three years.
- The power to the central unit should be shut off when the summer ends, otherwise the unit's heating elements could consume energy all winter long. Flip the circuit breaker if the unit doesn't have a separate switch. When the hot weather returns, be sure to turn the power back on at least 24 hours before starting it up again to avoid damaging the compressor.

Other tips:

- Keep all doors and windows closed when running air conditioners.
- Does your house really need to be at 68°F (20°C) all summer long? You'll save 3 percent to 5 percent on air conditioning costs for each degree you raise the thermostat.
- Seal unnecessary air conditioning ducts in the house with duct tape.
- To get rid of an old air conditioner, call your town's hazardous waste department. They will pick it up and drain the CFCs out of it before the unit is disposed of.

(*See also* Attics; Awnings and Sunshades; Ceiling Fans; Energy-Saving Tips.)

AIR PURIFICATION

You can buy an electronic air filter if you have serious allergies—some filters are freestanding, while others can be attached to the furnace. If you can't afford these expensive units, you might consider trying the following:

- Vacuum large pets on a regular basis. Cats can be trained to tolerate this if fed some catnip before the vacuuming. You might try the same method with a dog, bribing it with its favorite treat. (*See also* Pet Grooming.)
- Keep furniture and carpets free of cat and dog hairs by running a damp nylon toilet brush around the place. Or brush furniture lightly with a sponge dipped in a solution of warm water and glycerine.
- Don't let your vacuum bag fill up to more than three-quarters full. Dust will start filtering back into the air after that stage. You can place vacuum waste in the compost and reuse the bag.

- Make your own dusting oil: take three parts light mineral oil and mix with one part corn oil; add a drop of lemon or clove oil for a pleasant scent. Put this on your dust mop or dust rag and store in a used plastic bag. When the mop or rag becomes dirty, wash in warm water and reapply.
- If you have a ventilating fan in your kitchen, don't leave it on for any length of time. It can expel a lot of oxygen if the house is tightly sealed.
- Plants help scrub the air by filtering out dust and returning oxygen to the air. Spider plants are best for removing formaldehyde and other chemical irritants. Marginata and English ivy will also remove formaldehyde. Any 10- to 12-in (25- to 30-cm) potted plant will lower pollution in a 100-square-ft (9-m^2) area.

(*See also* Animal Odors; Cigarette Odor; Odors.)

ALUMINUM SIDING

Older aluminum siding sometimes becomes dull because of "chalking"—a mild form of corrosion. To remedy:

- Try washing with a pressure washer, using an environmentally friendly cleaning solution to remove dirt. If this doesn't do the trick, scrub the siding with a moderately stiff brush (you don't want to scratch) dipped in a solution of gentle detergent and water.
- After the surface is rinsed, allow the house to dry for several days, then apply one or two coats of a quality, 100% acrylic latex house paint. (If you have any leftover paint, make sure you dispose of it through your local toxic waste disposal facility. Never throw anything like this down the toilet or dump it in the backyard.)

To change the color of aluminum siding, begin by cleaning it using the above method. Then fill in any cracks or punctures with caulking compound (body filler for cars works, too). Rinse with clear water and apply an outdoor latex paint, available either in flat or semi-gloss finish.

(*See also* Painting: Outside.)

ANIMAL ODORS

- You can deodorize a dog by rubbing baking soda into its coat and then brushing it off. The baking soda will clean the coat as well.
- Oatmeal or baking soda can make a carpet smell clean and fresh: sprinkle around and then vacuum.
- If pets have left deposits around the house, wipe up, blot the area with rags and neutralize stain with vinegar, lemon juice or, if necessary, ammonia. Scrub area. Or moisten the area with cool water after removing feces and sponge with ½ tsp (2.5 mL) detergent and 1 tsp (5 mL) vinegar in 1 quart (1 L) warm water. Work this into the soiled area.
- Blot urine with rags. Then sponge thoroughly with cool water or with 1 cup (250 mL) white vinegar, followed by cool water. If odor remains, try a solution of 2 tbsp (30 mL) each of household ammonia and water.
- Clean kitty litter container with ½ cup (125 mL) vinegar.
- If your pet has had an encounter with a skunk, apply a coat of tomato juice. Wait until it darkens, then wash the animal. Wash repeatedly with a gentle shampoo; or if the animal isn't too big, let it have a good soak in concentrated citrus juice. After ten minutes, rinse thoroughly.
- Smelly pets could have a skin disease. If the odor persists, take them to a vet.

(*See also* Odors; Pet Grooming; Stain Removal: Fabrics.)

ANTS

Ants can be aggravating, but they are also beneficial. In addition to aerating soil, they prey on other pests, including flea and fly larvae, bed bugs and young silverfish. Many people can tolerate a few ants in their home, and there are plenty of steps you can take to minimize their numbers.

- Find their food supply and remove it. Food should be stored in tight containers. Keep garbage and compost materials in tight-lidded bins. Empty your garbage daily and clean your stovetop, counters and floors regularly.
- Locate their point of entry and seal it—temporarily with duct tape, and permanently with silicone caulking.
- A soapy sponge or a spray bottle filled with soapy water will kill individual ants and erase the chemical trail that the line of ants follows.

- Lure ants outside. If you place an attractive food source in a dirt-filled flower pot, ants will move in. Then place the pot outside and kill ants with hot or soapy water.
- Sprinkle sage around a cupboard which red ants have invaded.
- Place slices of lemon or sprinkle cinnamon where ants make an appearance.
- Soak a sponge in sugar-sweetened water and put it near where ants collect. Every so often plunge the whole thing in boiling water to kill off any ants.
- Put Epsom salts in the corners of each room, but not if you have small children who might be enticed by the possibility of an interesting new taste sensation. Pure Epsom salts taken internally can cause vomiting.
- Sprinkle borax on shelves.
- Plant mint near the house to discourage ants from entering. Onions have the same effect.
- To prevent ants from joining you on an outdoor picnic table, put the table legs in containers of water before setting out food. Ants are lousy swimmers.
- To get rid of ants around a hummingbird feeder hung on a pole, wrap rags soaked in cooking oil around the bottom of the pole.

(*See also* Carpenter Ants; Insect Control.)

APHIDS

These little pests suck the leaves, stems and fruit of plants. You can detect their presence by looking for cotton-batting-like material around buds and new leaves. Using pesticides will only kill off their natural predators. Try the following instead:

- Spray plants with a biodegradable insecticidal soap for several days. You'll need a direct hit on the insects to be effective.
- Plant companion plants—nasturtiums, garlic, chives, mint and petunias—to help keep aphids away. They will avoid roses planted with a companion garlic clove or placed near a chive plant.
- Wash off plants with a solution of mild soap and water.
- Use garlic and bug sprays.

(*See also* Insect Control.)

APPLIANCES: Cleaning

- Club soda will clean and shine at the same time.
- Quick cleanups: wash with watered-down household ammonia (one part ammonia to one part water).
- Remove yellow stains from appliances making a solution of ½ cup (125 mL) bleach and ¼ cup (60 mL) baking soda in 4 cups (1 L) of warm water. Wash the appliance and let sit for ten minutes. Rinse with clear water and dry.

(*See also* Cleaning Formulas; Kitchens: Cleaning and Maintenance; Refrigerators; Stoves and Ovens.)

ATTICS

- A reversible fan installed in the attic will bring in cool night air and expel warm daytime air. This will save energy if you have an air-conditioned house.
- To keep your house cool in summer and warm in winter, make sure that you not only have the attic properly insulated but that there is also at least 1 in (2.5 cm) left between the insulation and the underside of the roof.
- To ventilate an attic, you can also install vent louvers or plugs bought at the local hardware store.

AVOCADOS

- To keep an avocado from darkening when it's mashed up in a bowl, put the pit in with it until ready to use.
- To ripen an avocado, put it in a brown paper bag in a warm place.

AWNINGS AND SUNSHADES

These useful items can make your house a cool place in summer and eliminate the necessity of using an air conditioner. Carpets and furniture are also protected when you put up awnings over windows. Intense sun will fade carpets, damage furniture and ruin watercolors.

- Use awnings that are rigid and stay in one place only if you have a spot that gets sun most of the day, summer and winter. The more flexibility you have with your awnings, the better you will be able to control both light and temperature.
- Don't use light colors for awnings if you live near heavy traffic—they will get dirty very quickly.

- Be sure to dismantle awnings each winter so that sunlight can enter windows and help keep the house warm.
- When you store canvas awnings, make sure they are completely dry. Brush off any dirt or dust. Remove all stains. Cover in paper or an old sheet.
- Each spring spray with a canvas tent spray and let dry before reinstalling.
- Wash down aluminum awnings at least once a year.
- Vines make the most sensible sunshades of all. Quick-growing ones such as wisteria, kiwi vine (*Actinidia Chinensis Kolomitka*) and Virginia creeper (*Parthenocissus quinquefolia*) will provide shade during the summer and let sun in during the winter. They do not attract bugs.

BAGS

Paper or plastic? Each has its own problem. Paper bags use up trees, and pollutants are produced during their manufacture. Plastic consumes petroleum, creating pollution and contributing to the risk of oil spills.

- Instead, use cloth or string bags.
- Shop at stores where they will let you fill your own bags with groceries or where they will take back your old paper and plastic bags.
- Let the store owners in your area know that you'll give preferential treatment to anyone who promotes recycling.

Always try to recycle all your bags. Just by using the same bag twice, you reduce its impact on the environment by half.

- When packing, place shoes in a plastic bag to keep them separate from clean clothes. Bags can also hold dirty clothes over the course of a trip.
- A plastic bag can serve as an emergency rain hat.
- Kids will find dozens of uses for paper bags; they can paint on them, cut out masks, make puppets, etc.
- You can reuse paper bags by jotting down your grocery list or other household notes on them.
- As long as they're still clean after use, paper bags are excellent wrapping for packages and presents.

(*See also* Recycling.)

BANANAS

- To keep bananas from turning black in the fridge, put them in a plastic bag and keep in the crisper.
- To freeze extra bananas, mash, combine with one lemon to every six bananas and place in an airtight container. Use in baking.
- Never put bananas and apples in the same dish—apples give off ethylene gas which bananas just don't like—turns them brown very quickly.

BANISTERS

- To fix a wobbly banister, remove it and clean out the old glue. Use yellow carpenter's glue and screws to it re-attach it, fill any holes with wood putty and sand it neatly.
- A shaky newel post may be tightened by removing it, tightening the top nut on the rod holding it in place and gluing it.

BARBECUING

Barbecuing out of doors will help keep your house cool during the hottest summer months. It will save on electricity as well. There is some controversy about the carcinogenic properties of some barbecue briquettes. Make sure you buy those that don't have chemicals added to make them catch fire easily. Gas barbecues tend to be easier on the environment and won't create as much smoke to cloud the atmosphere.

- Brush grills with vegetable oil before you start and they'll stay cleaner.
- If you have an aluminum container around, you can reuse it by turning it into a drip pan.
- Clean the grill by covering it with a cookie sheet, shiny side down, and let both heat up for ten to fifteen minutes over the fire. The muck on the grill will drop right off.
- Or put grills in the laundry tub or bathtub and allow them to soak in water and ½ cup (125 mL) of household ammonia. Wash off with a brush and rinse thoroughly.
- You can reuse charcoal briquettes. Rather than letting the charcoal burn itself out, put the hot coals in an old pressure cooker. Cover and lock the top and the valve. The buildup of pressure will be contained by the valve. Warm air will escape and your charcoal will be intact. You can do this four times.

- Dental floss is the strongest material you can use to truss meat and it won't melt on the barbecue.
- Start food 4 to 6 in (10 to 15 cm) from the heat; the thicker the meat, the farther from the fire.
- Wood is ready when it's a mass of glowing embers; charcoal is ready when it is grayish-white.
- To barbecue fish, marinate it and then place it on a very hot grill and cover. Cook three to four minutes, then reposition at a 45° angle but don't flip. Cover again and cook another three to four minutes.

BASEMENTS

A cool, dry basement can make your whole house feel cool and clean.

- It takes a year to dry out new wood or concrete.
- By keeping a dehumidifier running in the basement all summer, you will keep your whole house cooler and drier. This may seem an extravagant waste of electricity, but perhaps it will eliminate the need for an air conditioner. (*See also* Dehumidifiers.)

Damp and Soggy Basement:

- Train fans or even a hair dryer on seriously damp spots. Then scrub concrete or cinder blocks with mild muriatic acid. Muriatic acid is toxic so proceed with caution. Follow with a latex masonry sealer paint.
- For constantly weeping walls, you'll have to find out where the moisture is coming from. It may be necessary to damp-proof from outside. (*See also* Concrete; Foundations.)

Floods:

Water is one of the best conductors of electricity, so if your basement floods, don't venture down there without wearing rubber boots and rubber gloves. If you should happen to lean against the wall with your feet in water, you'll be the perfect conductor.

- Use a wooden spoon to flick off the main switch or pull out the main fuse block.
- Shut off the water supply, especially if a ruptured water main has caused the flooding.

- For minor flooding, use mops and buckets to swab up the mess and empty outside, or rent a vacuum that will suck up the water.
- For a major flooding, rent a submersible electric pump. Connect your garden hose to the pump and run it outside away from the house, preferably to the nearest storm sewer. Make sure you plug in the pump outside the house or at a neighbor's.

Mildew:

- If it lingers, paint the walls with a lye solution of one 9½ oz (270 mL) can to a pail of water. Caution: Lye is caustic—handle it carefully.
- Charcoal, baking soda and lavender hung in muslin bags in offending areas will help absorb some of the smell and clean up the air a bit.

(*See also* Mildew.)

Odors:

Any sewer smell is definitely alien to your basement. Try the following:

- Pour boiling water down the drain and it will probably go away. If it doesn't, call a professional.

Waterproofing the Basement:

- Wrap cold water pipes with fiberglass insulation to prevent condensation from forming on them during humid weather.
- Keep roof gutters, drainpipes and basement drains clean. (*See also* Drains; Gutters.)
- Make sure water is draining away from the house. If it isn't, regrade the earth around the foundation so that it drains away properly.
- Seal all cracks in the basement walls with mortar. Do this from the outside if possible. It may be necessary to dig around the foundation. Cracked weeping tiles can be repaired or replaced, cracks filled and walls waterproofed. Such an undertaking offers an opportunity to add insulation.

BATHROOM SAFETY

- Don't keep pills and drugs of any sort in the bathroom. Keep them in a cool dry place well out of reach of small children.

The moisture isn't good for medications and children might get into them.

- Hot water can scald a child. Make sure the temperature on your water heater isn't set higher than 120°F (49°C). However, you'll have to check the minimum temperature your dishwasher needs to work efficiently if you do this.
- If you're in the bathtub and turn on a radio, you will be the perfect medium for a lethal electric shock. Any appliance dropped into water can be dangerous, even if it's plugged in but turned off. Electrical appliances in the bathroom should be used with great caution. Make sure the sockets are grounded.
- Put a lock on the outside of the bathroom door if you have children under the age of three.

(*See also* Childproofing Your Home.)

BATHROOMS: Cleaning and Maintenance

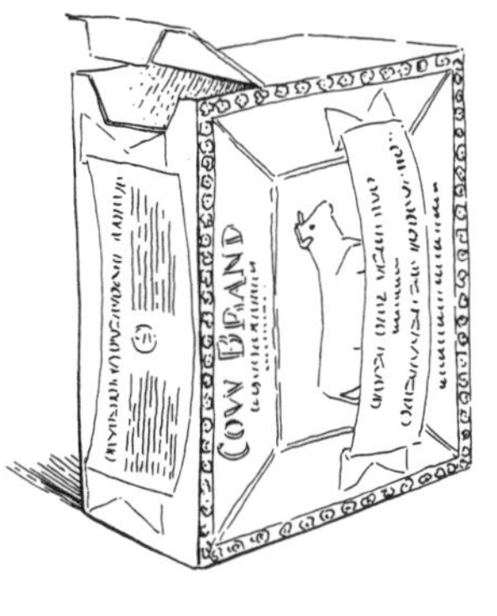

The simplest, safest and most cost-efficient cleaners can be made with a number of items you likely have in your home already. Always try an environmentally-safe option before turning to a chemical alternative.

Note: Be careful. A mixture of chlorine bleach and household ammonia creates toxic gas. Use one or the other, but never together. Ammonia, even on its own, produces strong fumes—make sure the bathroom is well ventilated.

- A 50:50 mixture of white vinegar and water is a good all-purpose cleaner that doesn't require rinsing. A stronger cleaner can be made by melting 1 tbsp (15 mL) soap flakes in 4 cups (1 L) of warm water, then adding 2 tbsp (30 mL) borax and 1 tsp (5 mL) lemon juice or white vinegar. Follow with a rinse.
- Baking soda and water is an excellent abrasive cleanser. Just rinse well after using.
- You can cut down that bathtub ring by putting a little baking soda in the water. Makes your skin feel good, too.
- To keep mirrors from steaming up, clean and then wipe down with a thin layer of glycerine.
- Leave shower doors slightly open to allow the shower area to dry after use. Keep shower curtains fully drawn to allow the moisture to dry on the inside.

- To prevent toilet bowl stains, use a toilet bowl scrub brush regularly. Just sprinkle in a little baking soda and brush around. (*See also* Toilets.)
- Very light bathroom stains on sink and tub surfaces will often come off by rubbing a chunk of lemon over them.
- Stained porcelain will brighten up with a paste made of cream of tartar and hydrogen peroxide. Let the paste sit for an hour, then lightly scrub it off.
- If porcelain is badly stained, try this method after everything else has failed: place layers of rags or cloths over the surface and pour bleach over them. Let sit for thirty minutes, then rinse.
- Here's a way of using old nail polish. Chipped porcelain can be disguised by applying thin layers of correction fluid and letting it dry between applications. Then cover with white nail polish to give the proper shiny surface. However, neither of these items is environmentally sound and should only be used if you want to get rid of them.

Bathtub Drains:

Rather than using the highly toxic commercial drain cleaners from your supermarket shelf, consider trying the following:

- 1 tbsp (15 mL) baking soda; 1 tsp (5 mL) salt; ½ cup (125 mL) vinegar. Use regularly to keep drains clear. (*See also* Drains.)

Cleaning Grout and Caulking:

- Grimy grout can be banished with a paste of baking soda and water, again using that old toothbrush. (*See also* Grout: Repair.)
- Terrible dark spots can be disguised with a bit of white toothpaste.
- To clean caulking, rub with an old toothbrush dipped in household bleach. Be careful not to let the bleach touch the tile; it can take off the finish. A rubber nonslip mat can be cleaned in the same way.

Showers:

- If a shower head isn't working efficiently, it could mean that there's a buildup of calcium or lime. Unscrew the head, put all the pieces in a bowl of vinegar to dissolve the deposits, then put it all back together again.
- Keep shower curtains fresh by putting them in the washing machine on a medium setting. Wash in warm water with a

combination of detergent and baking soda, ½ cup (125 mL) each. Throw in a couple of towels if you are washing plastic curtains. Add ½ cup (125 mL) of vinegar to the rinse water, but don't let the curtains spin dry. Hang them to dry instead.
- You can get rid of mildew on shower curtains with baking soda paste and water wiped off with lemon juice.
- Wipe down shower doors with a solution of equal parts of vinegar and water.
- Heavy film can be removed from shower doors with fine steel wool, then rinse with water.

Tiles:
- Just after you have showered and dressed is a good time to clean the shower tiles on a day set aside for housecleaning. The first part of the job is done—dampening the surface. Wash down with a solution of ½ cup (125 mL) of household ammonia, ½ cup (125 mL) vinegar and ¼ cup (60 mL) washing soda in a bucket of warm water.
- If this doesn't work make a paste of baking soda and bleach, then apply with a brush. Let it dry, then rinse off.
- If you have a buildup of soap and minerals caused by hard water, clean off tiles with a rag soaked in vinegar.
- Spray tiles and bathroom hardware with homemade polish. (*See also* Furniture Care and Maintenance: Furniture Polishes.) Pick up future spots with a wet cloth, then polish.

(*See also* Cleaning Formulas; Commercial Cleaning Products; Decals; Drains; Mildew; Toilets.)

BATHROOMS: Water Conservation
- Instead of a bath, take a five-minute shower.
- Install a low-flow shower head.
- When you brush your teeth, use a glass of water rather than leaving the tap running.
- If you need to install a new toilet, consider one of the new low-flush models.
- Standard toilets can flush away less water if you place a small, well-sealed container filled with stones, or something similarly heavy, in the tank.
- Don't use your toilet like a wastebasket.
- Fix toilets that run on after flushing.

(*See also* Water Conservation.)

BATHTUBS (See Bathrooms.)

BATS

Contrary to popular belief, these animals will not try to get into your hair and are wonderful as long as they are flying around outside and eating insect pests.

- If bats nest in your attic, seal off all openings except one and set a flashlight on the nest. They'll flee.
- When you're sure that the animals are out feeding, seal the remaining opening with mesh screening.
- Cut strips of foil from old pie plates and hang in front of entrances.
- If a bat flies into the house, don't panic. Turn off all the lights and open up doors and windows to let it escape.
- If you are going to disturb the nest anyway, or expect to be cleaning out the bat dung, be sure to wear a mask. The dung carries a disease that is sometimes fatal to humans.

BATTERIES

While household batteries make up a small proportion of the garbage in our landfills, they are a significant source of contaminants, particularly mercury.

- Used batteries can be incredibly toxic. Check with your local toxic waste depot for disposal instructions rather than throwing one in the garbage or abandoning it somewhere out of sight. The contents of the battery can leach into the soil.
- The most cost-effective and environmentally friendly batteries to use are the rechargeable ones. They last for about five years, cutting down considerably on your toxic garbage and adding to your pocketbook.
- Currently, the best rechargeables are nickel metal hydride batteries. They contain no toxic heavy metals, store more electricity, outlast other rechargeables and cost about the same. And when you need to dispose of them, they can be disposed of with your usual garbage.
- Keep your batteries in a cool, dry place. The shelf life of premium batteries is two to three years stored at room temperature. The hotter the storage area, the more quickly the batteries will deteriorate.

(*See also* Household Hazardous Products.)

BEADS: Restringing (See Jewelry.)

BEES

- To counteract a bee sting, put a piece of freshly sliced onion on top. Follow this with a water and baking soda paste.
- If bees are terrorizing your hummingbird feeder, put a guard of mesh screening over the feeding spout to keep them away from the nectar. (*See also* Wasps.)
- Remove the stinger by scraping it off your skin; do not try to grasp it as this will squeeze the stinger, releasing more venom into your system.
- If the person stung experiences shortness of breath, dizziness, or any unusual symptoms, get him or her to the nearest hospital immediately. Allergies to bee stings can have fatal results.

BICYCLE REPAIR

Using a bicycle can do your body good as well as cut down on the pollution in your neighborhood. Always walk or use a bicycle rather than a car for short-haul chores. But, like a car, the bicycle is not a machine you can just let go. Maintenance is important, otherwise it will become a pain not only on the bottom but on your pocketbook as well. A bicycle tool kit can consist of a set of small adjustable wrenches and a set of small locking pliers.

Cleaning your bike:

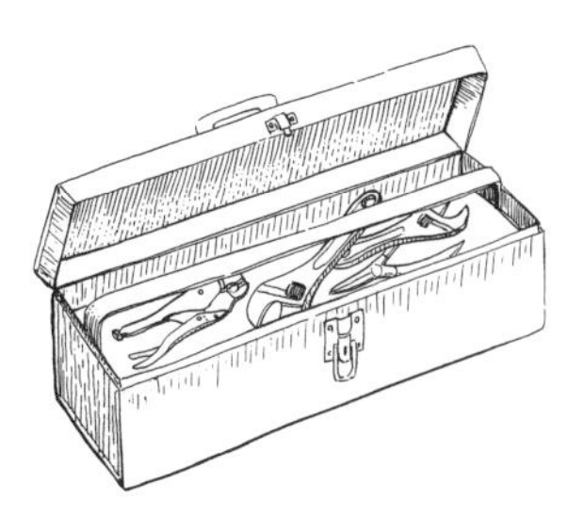

- Hose it down after long rides. If this is too much like work, do a serious job at least once a month.
- Be sure to keep the rims of the wheels clean at all times. Dirt and grease will pass from there to the brakes and ruin them.
- When you clean your bike, work from the top down with warm soapy water, followed by clear rinse water. But don't leave the water running—use a spray gun so the water's there only when you need it.
- Put the bike on a stand and remove the wheels to get at the area behind the hub and rims.
- If you have taped handlebars, apply soap and scrub with a nailbrush.
- Use an old toothbrush to clean the chains and the chainwheel teeth.

Lubricating Your Bike:

- Use lots of oil or grease when you do this mucky job. Be sure to wipe off any excess. It only attracts more dirt.
- White lithium grease or bike grease is good for: ball bearings, bottom bracket, brake cables, brake spring, seat post, stem, threads and wheel axle.
- Lubricant containing Teflon or silicone is good for: brake cables, brake pivot and derailleur pivots.
- Cycle oil or recycled motor oil is good for: interior of the free-wheel and the hub with internal gears.

BIRDS: As Pests

If you are plagued by pigeons, for instance, make sure that places under the eaves are kept clean so they aren't attracted to the area during nesting. Make banging noises at intervals if birds decide your balcony is a suitable place to land on a regular basis, or enclose the balcony with netting. Whatever you do, don't encourage them by feeding.

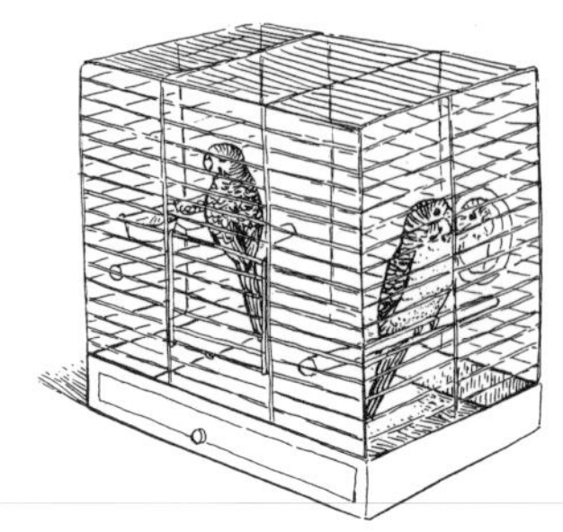

BIRDS: As Pets

Most birds need a companion and if you aren't prepared to devote a good deal of time to one, buy two. Canaries and budgies can live up to fifteen years. Make sure you buy birds that have been raised specifically as pets in captivity. Buying wild birds is cruel and can endanger species. Look for a very alert and sleek bird with bright eyes. Make sure it is at least six to eight weeks old.

Cages:

- Choose a cage that is large enough for the bird to get appropriate exercise. A two-cubic-ft (60-cm^3) cage is about the right size for a budgie or mynah bird. A parrot would need something larger.
- Install various sizes of perches and a swing. Bells and ladders can be introduced later, once the bird has settled in. Avoid toys like mirrors and plastic birds, as your pet may become more attached to them than to you.
- Place newspapers in layers cut to the size of the cage bottom. Remove the top layer each day. Put food and water dishes along the edge of the cage to make it easier to fill them.
- Make sure you keep the cage in only partial sunlight: most birds like to bask and then get cool. A 40-watt bulb on one side of the cage also works well.

- Provide a birdbath (a saucer full of water) at least once a week, or rinse a piece of parsley and attach it to the edge of the cage. The bird will peck and roll around on it to clean itself.

Feeding:

- Give a bird seeds daily, fresh green vegetables two or three times a week (limit lettuce to small amounts, however), a mineral block or food supplement daily and clean water daily. Treats can include bits of fruit, egg, cat chow or dry dog food.
- Place a cuttlebone in the cage to help keep the beak worn down and provide extra calcium. Put mineralized grit in a separate dish to help with your bird's digestion.

Health and Safety:

- Birds are happiest when they are able to leave their cage and fly. Your pet may have had its wings clipped at the store where you got it. Allow about six months for the feathers to grow back.
- Restrict the bird from flying into areas which may pose a danger. Kitchens pose several hazards, including hot burners and appliances, and pots or sinks filled with water. Avoid letting the bird out when company is over, as it could get trampled on or sat upon.
- While the bird is familiarizing itself with its flying space, keep window shades and draperies drawn, or place paper cut outs on windows to prevent the bird from flying into the glass.

How to Pick Up a Bird:

With one hand, quickly grasp the bird around the center of its body with your first and second fingers on either side of its neck.

BIRD WATCHING

- Put a simple bird-feeding station on your property and protect it from cats and squirrels with a metal shield or grease the pole to make it difficult to climb.
- If you have a hummingbird feeder, use one part sugar (no substitutes since they can cause a fungus in the bird's throat) to four parts water. Boil for two minutes. Adding red food coloring isn't necessary because the red finish on the feeder will attract the birds. Change the solution every three or four days. Wash the feeder with vinegar between feedings. Coat the wire on a hummingbird feeder with cooking oil to keep ants away.

- The best times to watch birds on your balcony or in a backyard is early morning or late afternoon.
- Keep the sun behind you, wear dull clothing and sit quietly so you can get a good look.
- Attract birds by putting colored marbles in the bird feeder.
- Leave lint and dog hair near a bird feeder for birds to use in nests.
- Attract birds by rolling pinecones first in bacon grease and then birdseed. Hang in trees or shrubs.

BLENDERS AND FOOD PROCESSORS

- Sluggish action may mean a buildup of sediment on the blade. Soak the jar and the blade mechanism in lukewarm water with dishwashing soap for eight hours. If this doesn't work, the blade bearing may be broken and will need to be replaced, or the drive stud is stripped, in which case you'll have to take it to a repair shop.
- If water splashes into the control panel, remove the panel and unscrew the protective plate underneath. Use an old toothbrush and scrub gently around the switch. Make sure the machine is unplugged and be careful not to disturb any wiring.
- To clean, half fill the container with hot water and a bit of dishwashing soap. Cover, turn on, then rinse and dry.
- When replacing your blender or processor, don't throw the old one away—donate it to charity. Even broken appliances can be useful for parts.

BLINDS (See Venetian Blinds.)

BOLT REMOVAL

- Always use the correct sized wrench to prevent damage.
- Use locking-up grip pliers or, in a pinch, an adjustable wrench.
- To get a recalcitrant nut moving, create a groove in the nut with a hammer and chisel—then pound with a rubber mallet—always counterclockwise.
- When a nut is frozen or rusted tight, apply a few drops of penetrating oil, which can be bought at a hardware store. Leave for half an hour and reapply if necessary.
- Rusty bolts can also be loosened with soda pop. Apply a cloth soaked in the pop and twist.

BOOKBINDING REPAIRS

- Fix a loose cover by using white glue applied with a knitting needle along the joints. Weigh it down for a day or so.
- Fix torn pages by putting waxed paper under the tear. Line up the rips and brush white glue on them. Cover with a light paper such as onion skin, rub and weight it down. Once the glue has dried, peel off any excess paper.
- Repair a loose page by cutting a strip of paper the length of the book, fold it lengthwise and apply white glue. Put one half on the loose page, the other on the following page. Cover the mend with waxed paper and weight it down. Let dry thoroughly.

BOOKCASES

Everyone knows about brick-and-board bookcases, but here's a variation:

- Use bricks with holes in them. Drill holes in the shelving and run a metal rod through them. It will line up the brick and boards and make them even steadier.

BOOTS

- An old laundry basket that is past its better days can be used as a good drip mat for boots. Cut down the rim.
- If your boots are soaked, dry the insides with a hair dryer set on moderate heat, or put your vacuum on exhaust and dry them with it.
- Boots dry better when they are standing up straight: put a roll of wire mesh in each one.
- Boot identification can be difficult for children. Help them find their boots by taping small strips of colored tape on each one.
- Use a colored clothes peg to clip your boots together.
- Salt stains can be removed by using a solution of equal parts of white vinegar and water on a damp cloth. Wipe clean.
- To improve the smell of your boots after a hard winter's use, stuff them with newspapers for a few weeks; the ink seems to absorb the smell.
- An ordinary pencil eraser will remove scuff marks from synthetic leather.
- If the toes of your boots are curling, remove the insole and cut a piece of #80 sandpaper 2 in (5 cm) wide the same length as the boot. Glue it coarse side up and replace the insole.

- Rub suede boots with stale bread; switch to a clean piece when the first piece gets dirty.

BOTTLED WATER

Bottled water has become a huge market. However, in the majority of cases, it is really a luxury not worth buying. It is expensive, and is primarily sold in plastic containers. Aside from that, not all bottled water comes from pure springs and glaciers like the images on the packaging would have you believe. A percentage of bottled water is just ordinary tap water that has had all the minerals removed.

- If you worry about the chlorine content of water or don't like the taste of it, fill a lidless container with tap water and leave it in the refrigerator overnight. The chlorine will evaporate.
- If your local water supply is contaminated, see if you can find water in glass bottles. You won't get that plastic taste, and you'll have a container that isn't made from petrochemicals.

BOW TIES

Cross one end over to the right about 1.5 in (3.8 cm) longer than the other end. Make a loop and begin to form the bow with the shorter end. Bring the longer end over and push it through the knot to form a completed bow. Make the ends even, then tighten.

BRASS

- For outdoor brass, polish using one of the methods below, then apply a light coat of paste wax.
- Scrub with salt combined with vinegar or lemon juice, but leave a fine film of lemon juice to keep the brass bright.
- Mix equal parts of salt and flour and add enough water to make a paste. Let dry, then rinse thoroughly and wipe with soft cloth. Use this method, and the next two, with caution—try a test patch first.
- Wearing rubber gloves, mix 1 tbsp (15 mL) household ammonia, 1 tbsp (15 mL) salt, 2 tbsp (30 mL) vinegar in one pint (half liter) hot water, rub on with a cloth, and polish with olive oil. All three methods require some vigor!
- The fourth variation is to combine equal parts of vinegar, salt and flour to form a paste. Paint on and let dry. Buff dry to a fine polish.

- If badly corroded, dip in a strong solution of washing soda and brush the corrosion with a nailbrush. Rinse. Dry and finish with metal polish.
- To get a high polish, rub with plain flour, which will remove any trace of old polish. Helps maintain shine as well.
- You can always paint lacquer (best is acrylic) on brass to protect it from tarnishing but never do this on andirons—the heat will cause them to peel.

BRASS BEDS

- To remove paint, take the bed to a professional furniture stripper.
- To save money, use a putty knife covered with a rag to scrape paint off.

BRICK CLEANING

To Clean Individual Bricks:

- Dampen brick and mortar thoroughly with a hose. This softens material and keeps down dust.
- Scrub with a wire brush and stack bricks neatly so they won't break.

To Clean Brick Walls or Patios:

- Use a wire brush and scrub clean.
- Wash down with garden hose or a pressure washer.
- Try removing stains from bricks by applying olive oil. Cover with rags or paper towels and leave on until the stains lift. It will probably take at least fifteen minutes. It would be wise to try a test area first. (*See also* Floors: Brick.)

BRUISES

- Plunge bruised area into cold water for thirty seconds to shrink surrounding blood vessels.
- Elevate the bruised area and squeeze edges with moderate pressure to prevent leakage into the injured area. Keep repeating this for fifteen minutes, depending on how severe the bruising is.
- Apply cold compresses to bruise, alternating with warm compresses of salt water.
- Apply a poultice. Soak clean pieces of cloth in a warm brew, squeeze out excess and apply to bruised area. Make the brew from comfrey root or powdered root if you can get it. Other poultices include plantain, chamomile, dandelion root and valerian.

BURGLAR ALARMS

Burglar alarms tend to be expensive, but they provide a great deal of comfort if you've ever had a break-in. There are two main kinds: perimeter (if a window or door is opened, an alarm sounds) and internal detector (an alarm sounds when sensors inside the house pick up movement or body heat). The best system combines both.

Burglar Alarm Tips:

- Hide the wires of perimeter systems under moldings and carpets, or behind walls.
- Make sure the control box is hidden in a closet and that the key switch or keypad is near the entrance.
- Put a "panic" button connected to the system near your bedroom door.
- Make sure the alarm or siren outside the house is well above the ground and not easy to get at.
- In areas that motion detectors don't scan, put down pressure-sensitive mats—they sound the alarm when stepped on. You may have some difficulty finding a company to install them, and they can be expensive, but in some situations, they're a good solution.
- Sirens should be put outside. Some companies now use strobe lights, too.

(*See also* Security: Houses.)

BURNS

Accident Prevention:

Burns are the most common accidents to young children. Follow these rules:

- Turn pot handles away from the edge of a stove.
- Don't keep treats anywhere near a stove.
- Try to use back burners for cooking.
- Don't use place mats that can be pulled from a table.
- Never, ever drink hot coffee or tea with a child in your lap.
- Turn the hot-water-tank thermostat down to 120°F (49°C). A temperature higher than that could burn a child. (Ensure your dishwasher will work efficiently at that temperature.)
- Start with cold water and then add hot water to a bath. When turning water off, turn the hot off first. Teach children to check the temperature of a bath first before stepping in.

- Don't leave hot water buckets unattended on the floor.
- Keep matches in airtight, childproof containers.
- Keep all smoking materials away from a child's reach.

First Aid for Minor Burns and Scalding:

- Immerse injured area immediately in cold water for ten minutes.
- Take off tight clothes and jewelry before swelling occurs.
- Put a piece of clean cotton or gauze over it to avoid infection. Never break blisters; this will retard the healing process and possibly produce scarring.

If Clothes Catch on Fire:

- Roll on the ground or smother flames with a towel or blanket (nothing synthetic).
- Once the flames are out, immerse burnt area in cold water for ten minutes.
- Do not try to pull off clothes stuck to the skin.
- Wrap the person in a clean sheet to keep bacteria from raw skin. Go to the hospital immediately.

(*See also* Childproofing Your Home.)

BUTTER

- Stale butter can be improved by kneading in cold water with a pinch of bicarbonate of soda and letting it soak for two hours.
- Stretch butter by whipping it: let it reach room temperature and beat with an electric beater until fluffy.
- Heat a knife to cut cold butter with ease.
- Clarified butter is made by heating butter over low heat and keeping it warm until the clear portion comes to the top.
- Freeze herbed butter in ice cube trays and you've got the basics for vegetable sauces readily available.

CAKES

- Stale cake or bread can be revived by brushing a little milk or cold water on the top and crisping in a medium oven for twenty minutes.
- If you want your cakes or any other baking to have a more sophisticated taste, use sweet rather than salted butter. It creams better as well.

(*See also* Thawing Food.)

CANDLES

- Candles will last longer if the wick is snipped close to the candle.
- Store candles in the freezer or soak them in salt water before lighting (dry them off before trying to light) to make them burn more slowly.
- Put elastic bands around the bottom of a candle that is too small for its holder.
- Put the base of a candle in boiling water if it's too large for the holder.
- Allow dripped wax to harden and cover with steamy towels. It should peel off easily.

(*See also* Wax: Candle.)

CANE

Bamboo:

- If the bamboo is soiled, scrub it with warm water and salt.

- If it's very dirty, wash with 3 tsp (15 mL) household ammonia and 3 tsp (15 mL) salt dissolved in 1 quart (1 L) water. Rinse with clear water and let dry. Finish off with furniture polish. (*See also* Furniture Care and Maintenance.)

Cane Chair Seat:

- If you have a sagging cane chair seat, turn upside down and sponge with very hot water. As it dries, the seat will shrink back to shape.

CANNING FOOD

- When canning fruit or vegetables, increase processing time by one minute for every 1,000 ft (300 m) you live above sea level. Add half a pound (1.1 kg) of pressure for each 1,000 ft (300 m) of elevation above sea level on a pressure canner.
- Low-acid foods such as vegetables (but not tomatoes), poultry, fish and meat should be processed in a pressure canner at 10 lb (4.5 kg) to a temperature of 250°F (116°C).
- High-acid food such as preserves, relishes and so on can be processed by the boiling water-bath method. They must boil for ten minutes at 200°F (93°C).
- Be sure to date jars so that you will know if there is any danger of spoilage through longevity.
- Normally I would suggest recycling old jars, but in the case of pressure canning, it's probably safer to buy new jars on sale. Use old jars for frozen jam and other preserves that don't require a lot of pressure.

CANS

- Never buy cans that have been dented, or are swollen or rusty.
- Most canned food will last safely for a year. Low-acid canned meat, poultry, stews and soup will keep from two to five years. High-acid contents (tomatoes, for example) will keep from twelve to eighteen months. If you see any deterioration of the can, get rid of it.
- If canned food has a funny smell, mold or foam on the surface—don't taste—toss. Botulism might be lurking in there and it's a killer.
- Store cans in a cool dry place where the temperature stays below 86°F (30°C); the shelf life of a can becomes very limited in temperatures over 100°F (38°C).

CARDIOPULMONARY RESUSCITATION
(See CPR.)

CARPENTER ANTS

These big black ants are more irritating than dangerous—unless they've really been at work in your house! They indicate that you might have nests (ergo decay) not only in your walls but also your furniture. Get rid of them and try to find their nests.

- Avoid attracting the ants by removing soft, moist wood near the house, repairing leaky roofs and walls, and removing damaged wood and debris near the house.
- Since carpenter ants can actually cause costly damage, you may need to use a commercial killer, like Wilson Liquid Antex. Ants like to eat it and take it back to the nest. It is deadly for the ants, their young, and the queen and without the queen, the colony dies off quickly.

CARPETS

Cleaning:

Looking after a carpet properly will add years of life to it. Try the following:

- Vacuum at least once a week with a vacuum cleaner attachment that will beat or stimulate the pile.
- To deodorize a carpet, sprinkle with baking soda or oatmeal before you vacuum.
- Damp salt sprinkled over the carpet will revive it like almost nothing else except baking soda. Rub in with a brush and let sit for several hours. Vacuum.
- Deal with spills and stains as quickly as possible. Begin at the outer edge and work to the center of the stain. Don't rub.
- Use vinegar and water on the browning caused by the efforts to remove the stain; spots can be removed with a dab of shaving cream (the nonspray type).
- For heavy spills, put a towel on the spot and cover it with something heavy. Replace the towel when it is damp.
- Use 2 tbsp (30 mL) vinegar or 2 tbsp (30 mL) detergent in 1 quart (1 L) of warm water. Gently work into stain and blot dry.
- Make a shampoo: whip washing powder, liquid detergent and water until suds hold their shape. Apply in the direction of the pile. Wipe up with a clean cloth.

- To clean an old carpet: vacuum and then sponge with the following solution: two parts water to one part vinegar with 1 tsp (5 mL) detergent for each 2 cups (500 mL) water. Let dry before walking on carpet.
- When you shampoo a carpet, put old plastic bags or tin foil around furniture legs and you won't have to move the furniture to prevent rust or water stains.
- To get rid of grease spots, sprinkle with cornstarch. Leave for an hour and brush clean. For difficult stains, blot with a mixture of vinegar and soapy water, or you may well have to resort to a solvent to lift the grease.
- If wax drips on the carpet, put a brown paper bag or paper towel over the mark and run a hot iron over it until the wax melts and sticks to the bag. Be especially careful doing this on nylon carpets—you don't want to melt the fibers.
- Pour cornmeal over mud; let dry and vacuum.
- Red wine spills: nothing works better than baking powder. Let dry and vacuum. Use white wine right away if you don't have enough baking powder.
- Chewing gum can be removed by putting an ice cube on it until it becomes brittle.
- To remove glue, try a cloth soaked in vinegar.
- Sometimes a clean art-gum eraser will pick up light footmarks.

(*See also* Paint Removal.)

Laying Carpets:

- When measuring for the amount of carpet you need, be sure to include window alcoves and other indentations; don't forget to allow for seams and remember that it must be laid in the same direction. Allow at least 3 in (7.5 cm) extra on all measurements so that the carpet can be fitted precisely.
- Try to avoid creating a seam in obvious areas or areas of high traffic. If you do create a seam, make sure it lies perpendicular to the main source of daylight in the room—rays of light will only accentuate the seam with a shadow.
- Always make sure the pile of separate pieces of carpet all point in the same direction.
- If there's a pattern in the carpet, you'll have to allow extra yardage so you can make the perfect match where necessary.
- To carpet stairs, order an extra 18 in (46 cm) so you can move it later on to disguise worn sections.
- Never join expensive woven carpets; get it done by a professional.

Repairing Carpets:

- Keep coasters underneath furniture legs to prevent indentations from forming. If one does form, lay a damp terry cloth towel over the spot and press with a steam iron (be careful not to melt nylon carpets). Once the towel dries, the indentation will disappear.
- If a large area of your carpet gets flattened out by a piece of furniture, revive it with a short hit from a steam iron (don't let it touch the carpet) or steamer, then brush.
- Keep an extra length of carpeting handy to make repairs for damage, such as burn holes. For small areas, snip out the burnt fibers with nail scissors, then pull new fibers from the extra piece of carpeting and secure them in the damaged area with rubber cement; wait until the glue dries before rearranging the fibers with a comb to suit the existing pile; trim if necessary.
- To patch a larger area: first lift up the carpet and draw a square on the carpet back larger than the damaged patch. Cut this square out of the carpet, using a sharp knife. Glue a patch of burlap on the back of the carpet to completely cover the square hole. Cut a new piece of carpet to match the hole. To prevent both the hole and the patch of carpeting from unravelling at the edges, apply a bit of rubber cement around the perimeter of both. Let it dry before applying the patch. Put strips of double-face carpet adhesive on the back of your patch and press it into place on the burlap backing.

CARS

Basic Car Care Tool Kit:

- set of combination wrenches from ⅜ in to 1 in (10 mm to 25 mm), ⅜ in (9 mm) drive socket wrench set including universal joint, spark plug socket (to fit your car's spark plugs), extensions
- Allen (hex) wrenches from .05 in to .25 in (1 mm to 6 mm)
- ball peen hammer
- offset ratchet screwdriver
- oil-filter wrench
- antifreeze hydrometer
- gooseneck funnel
- tire-pressure gauge
- 12-volt test light
- small wire brush
- battery jumper cables

- jacks (two for hoisting car and working underneath)
- spark plug gap gauge
- pliers
- vise grips
- torque wrench

An Emergency Kit Should Contain:
- belts (fan belts and other types)
- jack
- first aid kit
- work gloves
- ground cover
- rags and towels
- electrical tape
- flares
- solar blanket
- chain
- traction mats
- sandpaper (for battery)
- fire extinguisher
- snow shovel
- money
- flashlight
- white flag (emergency sign)
- name, address, phone number

Car Care:
- Keep a bag of biodegradable kitty litter or sand in the trunk—both are useful for getting out of icy situations.
- To keep trunk and doors from freezing, wipe the gaskets around them with vegetable oil. (*See also* Frozen Locks.)
- If you can't start the car in the morning, blow hot air from a hair dryer on the carburetor.
- Remove rust spots with a piece of crumpled foil or #0000 steel wool. Or try sponging off with brand-name colas. Boiling water works, too.
- Remove corrosion with club soda or baking soda and water.
- Before you go to a body shop, see if the bathroom plunger will pop out a dent in the car.
- Learn how to check battery and radiator fluid, the oil level and tire pressure of your car.
- Use antifreeze that does not contain ethylene glycol.

- Avoid parking your car in a heated garage. It will cause condensation in winter and speed up rusting.

Car Washing:
- Clean chrome by rubbing with aluminum foil dampened with cola.
- Wash your car with a biodegradable detergent. Try doing it in this order: spray all over with the hose to get the worst stuff off. Then work from the roof down, washing a section at a time with soap and water. Hose off and go to the next section. Dry with a soft cloth, then wax.
- Or consider taking your car to a car wash. Many recycle their water, ensuring that pollutants like grease and oil are filtered out safely rather than entering the water supply.
- Remove tar spots with raw linseed oil. Let stand until soft, then wipe off with the oil.

Car Windows:
- A wash with baking soda will keep grime off windshields and other glass and trim.
- Use a plastic net onion bag to wash insects off windows. It makes a good scrubber. Or wash with baking soda and water, then rinse well.
- Put rubber mats over windshields at night, so you won't have to scrape ice off them in the morning.
- To get rid of any hazy film, wipe with vinegar and rinse with water. Then buff with a soft cloth.

CARVING

Use a very sharp knife and a two-pronged carving fork with a thumb guard. The long, thin, flexible kind of knife with a round end is best for slicing. Use a shorter one with a point to cut off wings, legs and chops.

- It's a lot easier to carve most roasts if you let the meat stand for ten to fifteen minutes outside the oven.

Chicken and Turkey:
- Use dental floss for trussing fowl. It holds better than string.
- Start the carving by removing the drumsticks. If the drumsticks are large, slices of meat may be cut from them by carving parallel to the bone holding the bony end. Then remove the second joint and slice the meat from the thighbone. Next remove the wings and lastly slice the breast meat.

Ham or Leg of Lamb:

- Take a few slices lengthwise from the bottom so it sits steadily. Flip the roast over and slice perpendicular to the bone, then cut parallel to the bone to remove the slices.

Roast Beef:

- Rib roasts: slice across the top, then vertically along the ribs to detach the slices.

CASSETTE TAPES

- If a tape becomes snarled in a cassette player, pull it out gently with a crochet hook or a large paper clip twisted into a hook until the snag is no longer inside the cassette. Straighten out the tape, then use the eraser end of a pencil to slowly rewind the reel, feeding the unsnarled tape back into the cassette.
- Clean tape heads with cotton swabs dipped in denatured alcohol.
- A broken tape can be fixed with a tiny dab of clear nail polish. (This is a good way to use up old nail polish.)

CAST IRON CARE

- If you have one of those old-fashioned cast iron stoves, clean it with stove polish or protect it with a thin film of mineral or vegetable oil.
- Be sure to cure a cast iron pot to prevent rust and ensure food won't stick. To cure: cover the bottom with cooking oil and put the pot over high heat until the oil smokes. Cook for about ten minutes, then cool and wipe off excess oil.
- Scrub out a cast iron pot with salt if it's giving food a funny taste.
- Never clean cast iron with soap or detergent. Wipe it clean with a damp cloth.

(*See also* Pots and Pans.)

CATS

Your cat can stay alone for forty-eight hours if you leave enough food and water plus a few toys: a big paper bag, a ping-pong ball to knock around and a piece of crumpled up paper suspended from the back of a chair. Cats spend two-thirds of their lives sleeping anyway.

- If your cat urinates on a new mate's side of the bed or in his or her shoes, it could be jealous.
- Vets recommend that cats be kept housebound. Good luck.
- If a cat urinates in the house, rub the spot with household ammonia and the cat won't do it there again.
- Never clean up cat vomit immediately—let it dry and it will slide up a knife without leaving a trace.
- Try the following cat exercises with your cat. Attach a ball of crushed paper to a piece of string. Tie it to your foot. Practice kicking so that the cat jumps up after it. It will give you both a good stretch. Or tie a piece of string with a soft ball at the end to a coat hanger hung high enough for your pet so it can enjoy batting it around.
- Buy biodegradable kitty litter. Natural litters break down in landfill or can even be composted and used as mulch as long as they're thoroughly scooped.

Feeding Tips:

- Kittens under three months should be fed four times a day; at three months, three times a day; at five months and from then on, twice a day. Cats are carnivores and need high protein food.
- Don't feed a cat table scraps. They are not a balanced diet.
- If your cat has diarrhea, it may mean it can't digest milk—remove it from its diet.
- Give your cat a sardine or two once a month to keep its fur from balling. It relieves constipation as well.
- Mash up some squash and mix it in with food (even if your cat is on a dry diet). Add a bit of salt to keep it drinking. Good for helping to pass stools.
- Always have fresh water available. If you have a male cat, make sure the water is pure. Salt and chemical-filled water can lead to urinary tract disease causing kidney failure. All cats should be fed low-ash food for the same reason.

Health Care for Cats:

- Invest in annual feline leukemia shots.

(*See also* Pet Grooming.)

CAULKING

- Caulking helps save energy. To find out where you need to caulk in your house, look for cracks in existing caulking around windows and doors and where different materials join—the chimney brick to the roof flashing, for instance.
- Interior caulking needs can be detected by holding a candle near window and door frames and wherever you might suspect a draft.
- Never caulk in temperatures lower than 50°F (10°C), otherwise the caulking material won't adhere properly.
- Use the long twist or rope caulking to create temporary seals around storm windows. It can be removed later. Push the rope caulking into place with your thumb.
- Put masking tape all the way around where you want to caulk to make a clean line. After you've put the caulk down, dip your finger in water and trace the outline of the caulking. Remove tape before it dries.
- If you find gaps that are too big to fill with caulking, stuff them with insulation or weather stripping.
- Caulk the places where telephone, cable and other wires come into the house.
- Caulk along baseboards if you feel a draft running along the floor.

CEILING FANS

- Fans are cheaper and more environmentally friendly than an air conditioner. If a ceiling fan is placed at the top of a stairway, or if you have a cathedral ceiling, in the winter it will push the warm air where it's needed—lower down.
- Never install a fan on a low ceiling. It must be at least 10 ft (3 m) above the floor. It's not a good idea to put one over a dining room table either. Apart from cooling off food and making candles gutter, people tend to lean forward while standing to push away from the table. It could be dangerous.
- Use a feather duster on a broom handle and dust on a weekly basis. Then you won't have to do the following.
- To clean ceiling fans, wipe all the accumulated grease off with a damp cloth. Too much water on the blades could warp them. If the grease is particularly thick, follow up with a scrub with general cleaning formula. (*See also* Cleaning Formulas.)

CENTIPEDES AND MILLIPEDES

- Sprinkle borax around windows, doors and any other place these creatures might enter. Caution—borax should not be placed where children or pets could find it and eat it.

(*See also* Insect Control.)

CERAMIC TILES

- When installing, take a piece of string the length of the first row of tiles and rub it with colored chalk. Thumbtack the string at each end in a taut line at exactly the height of a tile, then snap the string so that it puffs out a chalk line on to the wall as a guide for installing the first row of tiles. Much better than using a wooden yardstick.
- Start laying tiles at the center of your room, floor or counter. Find the center, then set a vertical and horizontal, absolutely square chalk line. It's a good idea to test-fit the tiles to see how the edge tiles will have to be cut.
- Apply only a small bit of adhesive at a time so it doesn't harden. Spread the mastic with a special mastic tool, set the tiles and spacers and let it set for a day before moving on to grouting.
- To cut partial tiles, use a china-marking pencil to draw a line where you want to cut the tile and score this line with a glass cutter. Place the tile over a straight object (a board or tabletop) and press evenly on either side of the scored line. The tile will snap along the line.
- To replace a broken tile, cut away the damaged tile and make sure the surface is safely waterproofed. Apply new tile with tile adhesive. Fill with a grout cement.

Cleaning Tile Floors:

- Add 1 cup (250 mL) vinegar to 2 gallons (9 L) of water and sponge mop.
- Clean ceramic tiles with soap and water and a rubber scraper—the kind they use at gas stations to clean your windows.

(*See also* Cleaning Formulas.)

CHANDELIERS (See Lighting.)

CHEESE

- Make a low-calorie cheese by putting plain yogurt into a muslin-lined sieve or bag and letting it drip overnight into a bowl.
- Always serve cheese at room temperature. Take it out of the fridge about an hour before you intend to eat it.
- You can successfully freeze the following cheeses: cheddar, Swiss, and all French, Italian and Greek natural cheeses. (*See also* Thawing Food.)
- If a hard cheese develops some moldy spots, cut them out—the quality of the cheese won't suffer.

Grated Cheese:

- Always buy chunks of Parmesan and other cheeses for grating. It's much better when you make your own and less likely to be stale because you can chop off only what you need. Cheddar and Parmesan both freeze well, so look for bargains.
- Cheese equivalents: 2 oz (57 g) cheese equals ½ cup (125 mL) grated cheese; 1 lb (500 g) cheese equals 4 cups (1 L) grated.
- To ease grating, put a drop of vegetable oil on the surface of the grater. It will wash up much more rapidly as well.
- If you want to grate a very dry cheese, cut it into small pieces and use a blender or food processor.
- To make a really good cheese topping, grate Parmesan and a herb such as parsley, along with some bread, in the food processor and sprinkle over a casserole.

Storing Cheese:

- The best way to store cheese is to keep it in its wrapper. This will probably be an environmentally unfriendly plastic bag. Reuse them for as long as possible. Air cheese on a regular basis to keep it fresh.

CHEWING GUM REMOVAL

- Gum on a scarf or shoe can be removed by putting the item in the freezer for a couple of hours then scraping the gum off.
- Gum ground into the carpet can be chilled with ice cubes wrapped in a plastic bag. Lift off gently.
- Remove chewing gum from hair with cold cream or a little peanut butter. Freeze with an ice cube and then lift off.

- Or soak a rag in eucalyptus oil and sponge off.
- Put egg white over the gum and let dry. Then lever it off with a knife.

(*See also* Stain Removal: Fabrics.)

CHILDPROOFING YOUR HOME

- Keep stairs well lit, and if they're carpeted, make sure that there are no worn spots. Don't let children play on stairs. Be sure to install a sturdy stair gate to keep small children away from the area.
- Store area rugs until children are old enough not to slip on them.
- Use plug outlet covers painted the same color as the wall on all unused outlets.
- Replace glass-topped tables with safety glass.
- Put self-adhesive weather stripping around the very sharp edges of furniture.
- Make sure fireplaces are well covered. Glass fireplace doors can become hot. Try blocking them off to children or avoid using the fireplace when they're wandering around.
- Install safety locks on all windows.
- Make sure window screens are solidly installed.
- Never place a piece of furniture underneath a window. A child may be tempted to crawl up on it.
- Anchor bookshelves to the floor.
- Place heavy furniture in front of outlets. Put any floor lamps out of the way behind sofas.
- Make lots of wide open spaces between furniture for playing. Put all precious items higher than 4 ft (1 m).
- Keep the washer and dryer locked, or the area where they are kept locked and out of bounds.
- Anything lethal—chemicals, paint, detergent or gasoline—should be properly disposed of when you have children around. Or they should be stored in a high, locked cupboard that has good ventilation.
- Make sure you don't have any poisonous plants around. (*See also* Poisonous Plants.)
- Don't let toddlers near kitty litter boxes. They'll try to eat anything. This includes garbage—keep the container latched.
- Tape lamp cords around table legs to keep lamps from being pulled off tables.
- Put colored tape on sliding or glass doors at the child's eye level.

- To keep track of what doors your child is using, or sneaking out of, attach bells to each one—out of reach of little hands.
- Don't hang shopping bags from a stroller or carriage—it could tip it over.

Bathroom:

- Keep all medication stored safely in your bedroom—definitely not in the bathroom cabinet.
- Mark the hot-water taps all through the house with bright red decals or red paint.
- If your children have reached the age when they like to bathe themselves, put a decal in the bath to indicate a safe water level. And make sure there's a nonslip mat in the tub.
- Never leave a young child alone in the tub.
- Keep cosmetics, perfume and deodorants out of reach: a basket hanging from the ceiling, perhaps?
- Never have a heater in the bathroom.
- Put a safety- or night-light in the razor socket.
- Have a lock on the bathroom door that can be opened from the outside. Keep a latch on the bathroom door to keep little ones out—they can drown in toilet bowls.
- Always have a covered diaper pail with a childproof lock—add vinegar to the water to deodorize.

Practice these safety measures when bathing a small child:

- Use an infant seat for baths. Put a towel on the seat.
- A plastic clothes basket lined with a towel and set in the tub with a few inches of water will contain a small child.
- Never tie swinging toys to cribs or playpens—a baby can strangle on the string.

Kitchen:

- Remove control knobs on stove when not in use.
- Keep freezer locked. Always remove a freezer, fridge or oven door before discarding it.
- Choose a dishwasher with an interior lock.
- Keep children away from detergent—it's toxic.
- Unplug appliances and keep them away from the sink—electricity and water can kill.
- Latch drawers and cupboards containing knives or breakables with rubber bands or catches bought at the hardware store.
- Keep cleaning products out of reach and away from food.

- Lock up plastic bags of any size.
- Don't use aerosol sprays. They aren't good for the ozone layer anyway. Even those with Ozone Friendly on them can blow up under pressure.
- Keep glass bottles of pop in a cool place where they can't be tipped on the floor. They could explode.

Nursery:
- Make sure that all baby and children's furniture is painted with unleaded paint.
- Make sure the bars on the crib are close enough together—no wider than 2¾ in (7 cm).
- Use bumper pads that are firmly tied in place, and never place a crib or playpen near a window, fireplace or radiator.
- If you use a mesh playpen, buy the kind without a drop side. If it collapses, a child could suffocate.
- Put a mattress next to the bed when your child moves from crib to bed. If you haven't an extra mattress, use pillows or a couple of blankets.

Toys:
- Always test toys yourself before handing them over to a small child.
- Keep stuffed toys clean with cornstarch—just rub in and brush off.

(*See also* Burns; Internet Safety; Play Safety; Transporting Children.)

CHIMNEYS: Care and Cleaning
- Inspect your chimney annually: check for loose bricks in the hearth and loose or cracked brick on the outside of the chimney. Always make sure the flue is clear.
- To clean a chimney, the easiest and most practical thing to do is to hire a chimney sweep—yes, they still exist. If you have an older house, it would be safer to have a professional do the cleaning at once to ensure its safety and good condition.
- If you do decide to clean it yourself and if your chimney isn't an insulated metal one or hasn't a cast-in-place liner, rent or buy a long-handled wire brush from a fireplace or wood stove shop.
- Block off the flue opening or openings (there may be more than one) with cardboard taped into place. Soot will fall to the bottom. After the job is done, you can take the tape off and

vacuum up the sooty residue. Take care to brush down the smoke chamber before you vacuum. You must vacuum behind the damper, as well.

- You can keep birds from building nests in your chimney by fitting a wire cage or a box made of wire mesh over the top of the chimney. This will keep sparks from flying on to your roof, too.
- Even if you don't have a working fireplace, you might have a chimney that vents the furnace. Hold a mirror to the bottom opening and if your chimney is straight and you don't see sky, you've got a problem that should be looked at immediately or tended to as above.

CHINA

- To prevent chinaware from getting chipped, stack plates with cloth napkins in between each one. Hang cups from hooks instead of stacking them.
- If your most precious and oldest china has developed yellow stains, wet a soft cloth, dip it in baking soda or a combination of lemon juice and salt, and clean. Only if the stains seem impossible to remove, try hydrogen peroxide.
- Any crack that becomes stained or yellow can be cleaned with baking powder. Leave it on for two hours then wash and wipe as usual.
- If china gets dull, rub olive oil over the surface and leave for an hour. Polish with a clean cloth.

Mending China:

Don't throw out precious bits of china, porcelain or glass. They can be fixed if you use these tips:

- Apply a very thin layer of epoxy after cleaning each of the surfaces to be joined. Try taping the pieces together with masking tape. Or clamp and support. Let dry for eight hours.
- After the pieces are clamped together, more glue will ooze out. Simply wipe with cotton batting and a solvent. If you happen to have some old nail polish remover still in your house, you can get rid of it this way.

Supports:

- Plates: set the plate in a piece of Plasticine in a drawer. Close drawer to hold upright. Or make a wax mold of the unbroken half of the plate with melted paraffin. Let set and move to broken side of plate. Fit first piece in and, as it sets, add others, gluing one edge each time. Or make a mold of the unbroken half with modeling clay, then put the pieces inside the mold to dry.

- Cup: keep the glued handle in place with strips of masking tape. Support the cup in a softened lump of Plasticine.
- Glass: put sticks with Plasticine tips in cans of sand on three sides. Prop glass in this.

CHOKING

Start with prevention in small children. Train them never to run with food or any object in their mouths. Make sure that food is cut into small pieces and chewed properly before swallowing. Always keep your eye on an infant and keep small objects, including peanuts, away from him or her. If choking takes place, call your local emergency number and:

- Straddle an infant over your arm. Support the head by holding the jaw and keep the head lower than the rest of the body. With the heel of an open hand, make four blows between the baby's shoulder blades.
- Lay a child over your knee with the head down. Give a sharp slap between the shoulder blades to dislodge the object. Don't try to scoop any object out of a child's mouth if it's at the back of the throat.
- The Heimlich maneuver can be used on a conscious person who is standing or sitting. This is done by standing behind the victim, wrapping your arms around his or her waist. Make a fist and roll it up against the abdomen between the navel and the ziphoid (the tip between the ribs). Grab your fist and press up quickly to dislodge any obstruction. Do this until it's relieved.
- If the child or adult stops breathing, use cardiopulmonary resuscitation. (*See also* CPR.)

CHRISTMAS TREES

- If you buy a fresh Christmas tree, make sure that it's from a tree farm and not cut down in the wild.
- A Christmas tree becomes a fire hazard the minute it dries out. Since a tree is cut many days before you actually buy it, keep it outside for as long as possible.

- Always buy the freshest one you can find—the needles should still be pliable, which indicates good moisture content.
- Look for a bright, green color. This will indicate a reasonably fresh-cut tree. But don't fall for something that looks like dye has been sprayed on or added to the water.
- To get it home, wrap the tree in burlap or an old sheet if you tie it to the roof of the car. Leaving it exposed will strip its needles faster than anything else. Point the tip to the rear of the car.
- Fireproof the tree by combining ½ gallon (2 L) lukewarm water, 4 oz (113 mL) boric acid, 1 cup (250 mL) alum and 2 tbsp (30 mL) borax. After it dissolves add 1½ gallon (7 L) water. Spray the tree and all your other Christmas decorations.
- Saw off at least an inch (2.5 cm) of the base, just as you would fresh-cut flowers. Put the tree in a bucket of the following dissolved in hot water: 2 cups (500 mL) corn syrup, 4 tsp (20 mL) bleach and 4 tbsp (60 mL) micronized iron (available from a garden supply center).
- Never place the tree near a fireplace or other source of heat.
- Put the tree and stand on something waterproof—an old piece of linoleum would be ideal—to prevent water stains on the flooring.
- If your community composts Christmas trees, first remove all decorations—even tinsel. Throw the tree out with a minimum of spilled needles by wrapping it in an old bed sheet before heading out the door. Or chop up the tree yourself and recycle it as mulch.
- If you decide to buy an artificial tree, make sure it is a simple but elegant one—that way it can be passed down from generation to generation.
- Store Christmas ornaments in egg cartons.

CHROME

- Rub with a pencil eraser to both clean and polish.
- Or rub with fine steel wool.

CIGARETTE BURNS

- Hardwood floor: if your floor is finished with urethane, try to wipe away only the burn very carefully with very fine sandpaper; if it's badly burned, you may have to remove the piece. If the floor has an oil and wax finish, sand the burn out, re-stain, and in a day or two, buff with wax.
- With a deep burn, fill the hole with wood filler after sanding. Sand again and apply two coats of varnish.

- Vinyl floor: cut out the burned area and insert a patch of the same material.
- Carpets: rub around the burn with steel wool or sandpaper to lift the fibers. Mix up ¼ cup (60 mL) borax in 2 cups (500 mL) hot water and dampen the spot with a cloth. Sponge with clear water. This also works on upholstery.

(*See also* Carpets; Furniture Repairs.)

CIGARETTE ODOR

Ex-smokers have the most sensitive noses around. If you've recently quit smoking, be sure you air all closets with a fan and have your carpets and broadloom shampooed.

- Wash the clothes you wear most often. Clothes hold the smell for weeks.
- A dish of vinegar will help if you don't mind the masking odor.
- During a party, burn candles to keep down the smell.
- Put a little pot pourri into a pan of water along with a stick of cinnamon and let simmer.

(*See also* Air Purification; Odors.)

CLEANING FORMULAS

Be careful when you are mixing your own cleaners: ammonia and chlorine will produce deadly fumes; and never mix chlorine bleach with vinegar, which creates potentially deadly gas.

All-Purpose Liquid Cleaners:

- Mix equal parts water and white vinegar. Simply wipe surfaces—no rinsing is required.
- Melt 1 tbsp (15 mL) soap flakes in 4 cups (1 L) of warm or hot water. This solution can be made stronger by adding 2 tbsp (30 mL) borax and 1 tsp (5 mL) lemon juice or white vinegar. Apply to the surface and rinse.
- Or combine ½ cup (125 mL) ammonia, ½ cup (125 mL) white vinegar, ¼ cup (60 mL) baking soda and ½ gal (2 L) water. Mix thoroughly.

All-Purpose Abrasive Cleaner:

- A gentle abrasive is a mixture of baking soda and water. This can safely clean most surfaces. Rinse well.

- A stronger abrasive is two parts borax and one part baking soda or washing soda. Sprinkle onto the surface, scrub with a damp cloth and wipe dry.

Absorbing Powders:
These are excellent cleaners. There are four basic types: talc, cornstarch, fuller's earth and baking soda.

- Talc and cornstarch are best for removing wet spots on carpets. They are so dry that they literally suck up moisture.
- Fuller's earth is available in either white or brown. It has the properties of clay and is superb for removing grease from materials such as clothing or carpeting. Sprinkle powder on spot, leave it for a couple of hours, scrape away the light crust that forms. For greasy areas, apply, leave for a couple of hours and then vacuum.
- Baking soda is an excellent deodorizer: sprinkle on carpets, leave for an hour and then vacuum. Sprinkle the bottom of the kitty litter box to absorb urine odor. Leave open in the refrigerator or freezer to absorb odors.

Bleach Alternative:
- Try a mixture of borax and water.

Cleaning and Polishing:
- To be ecological in your cleaning and polishing, always use rags and cloths rather than expensive paper towels, which not only destroy trees but use acid in the milling process. If you must use paper towels occasionally, buy those milled with recycled paper.
- Clean any metal with mild soap and water then dry with a soft cloth.
- For more difficult stains, make a paste of equal parts of salt, vinegar and flour. Use a soft cloth to apply, then rinse with clear water. Buff with another soft, clean cloth.

(*See also* Commercial Cleaning Products.)

CLOSETS
A sloppy closet reflects a sloppy mind, or so we're told. You can have a neat one by following these instructions:

Cleaning Closet:

- Place all the things you use most often between waist and eye level. Don't place heavy items you use often, such as the vacuum cleaner, behind less-used items. Put seldom-used objects higher in the closet, but not out of reach—for most people this is 10 in (25 cm) above their heads.
- Make a hanging board for brooms and other long-handled household equipment by creating a hanging board—hammer nails into a board at intervals wide enough to hold the heads of the equipment and keep them upright. Attach the board to the wall at about 7 ft (2 m) level.
- Deep corner cupboards in a kitchen can be made more useful by installing a lazy Susan. You might also consider having a piano-hinged door installed—this allows the entire corner space to be revealed.

Clothes Closet:

- Visit a closet-organizer store and steal some of their good ideas.
- Install adjustable shelves that are easy to move around to suit changing storage needs.
- Install two rods in the space of one. Shirts and blouses above; pants and skirts below.
- Leave about 1 ft (30 cm) of clear space with a higher rod for longer dresses and coats.
- If a cupboard or drawer is too deep to be useful, install a basket that can be lifted to reveal all its contents.
- Dark closets are difficult to keep organized, so make sure they are well lit.
- Potpourri and proper ventilation keep a closet fresh.
- Replace closet doors with louvered doors to increase ventilation.

CLOTHES CLEANING (See Clothes Washing; Commercial Cleaning Products; Dry Cleaning; Hand-Washing Clothes; Leather; Stain Removal: Fabrics; Suede; Velvet.)

CLOTHES DRYING

Clothes dryers are heavy users of electricity. Use them judiciously, mainly when you have a full load, and don't let them run through the full cycle—it overdries clothing. If you don't live in an area of major fallout from air pollution, consider drying clothes out of doors. There are handy fold-up gadgets that hold a large number of items. These are also excellent for drying

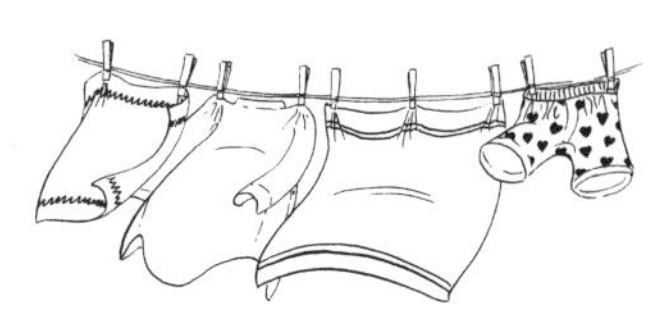

sweaters and delicate clothing that need to be laid flat. Always wipe down first with a cloth if you store it outside. Putting up an old-fashioned clothesline to use in fine weather will save you money as well.

When you use the dryer, use it only during off hours (usually late at night) and try the following:

- When taking clothes from the washer, shake each item loose before placing in the dryer.
- Sort according to weight—no heavy towels with delicate panties.
- Dry three to five shirts or blouses at a time for only five minutes. Then hang to finish drying. It will almost eliminate ironing. Crowding leaves wrinkles and uses up electricity.
- Don't let clothes get completely dry. Shake and hang clothes as soon as most of the moisture is gone and they have lost wrinkles.
- Never machine dry silk or wool.
- Remove lint from the machine at least twice during a heavy cycle.
- If the lint in the filter is damp, check the outside vent—it may be clogged.
- Netting put in the dryer will catch lint.
- Dry corduroys very slowly and while still damp, brush them off.
- Small hand-washable clothes can be dried with a spin in the salad spinner.

(*See also* Dryers.)

CLOTHES PACKING (See Packing Clothes.)

CLOTHES REPAIRING (See Darning; Mending Clothes; Patching Clothes.)

CLOTHES STORAGE

- Do not store clothes in air-tight plastic. Wrap clothes in cotton or in acid-free paper so they'll breathe.
- Washing clothes before storing will discourage moths.
- Air and brush wool after wearing.
- Use padded or wooden hangers. Wire leaves marks.
- Be sure everything in the closet is worn regularly. Store the rest.
- Store similar items together: belts, scarves, jewelry, shoes, shirts, dresses, etc.
- To protect against mildew, never put damp clothes in storage. (*See also* Mildew.)

- Good alternatives to mothballs are cedar chips, or dried lavender, rosemary or mint. Or try sprinkling clothes with pepper. (*See also* Moths.)

Boxes:
- Use boxes of the same size for stacking. Label carefully. Make sure they are well ventilated and waterproof.

Garment Bags:
- Use cloth garment bags only.

Cedar Chests and Closets:
- Line with acid-free paper to protect clothes from possible resins.

Mothball Odor:
- To remove mothball odor, leave a pot of kitty litter in the area overnight.
- Herbal sachets will help disguise mothball odor. To make herbal sachets combine 1 oz (30 g) of each of the following—bay leaf, lavender, rosemary, cedar chips and patchouli with a pinch of clove. Divide into bags. For a drawer use one bag, for a small closet, two bags, and for a walk-in closet, four bags.
- Put whole cloves in pockets of coats or sweater bags.

Trunks and Suitcases:
- Put clothes in a pillowcase before storing for easy removal when you travel.

CLOTHES WASHING

General Information:
- Keep your washing machine clean by adding a ½ gallon (2 L) vinegar and let the machine run through a whole cycle. Do this once a year.
- To keep detergent at its most efficient, store it in a plastic bucket with an air-tight top.
- Wash cycles should be only eight minutes long or the dirt makes a re-entry into the clothes.

A Few Tips about Sorting:
- Don't mix whites and colors. Never wash clothes made of flame-retardant material with soap—it destroys the finish.

Always mix small and large items in each load for maximum cleaning power.
- Put lingerie in a net bag or pillowcase fastened with a tie or closed by a zipper to prevent tangling. Use this method for keeping socks or any other small items from getting lost.
- Pin socks together and one won't mysteriously disappear.

Alternatives to Detergents and Other Laundry Products:
- When you make the switch from commercial detergents to homemade cleaners, you will first have to get rid of the detergent residue in all your clothes, otherwise yellowing may occur. Wash each load with ¼ cup (60 mL) of washing soda and water as hot as the product can stand.
- All-purpose soap: Combine 1 cup (250 mL) soap flakes, ½ cup (125 mL) borax and ½ cup (125 mL) washing soda. Store in a properly labeled container. For each load, add ½ cup of the mixture to the water in the washing machine before adding clothes.
- Soap for delicate items: Mix ½ cup (125 mL) borax, 1 cup (250 mL) soap flakes and 4 cups (1 L) of boiling water. Once cooled, store in a labeled bottle. Use one-quarter of the solution for each 4 cups (1 L) of washing water. Add clothes and wash gently, followed by a thorough rinse.
- Starch: Mix together 1 tbsp (15 mL) cornstarch with 1 cup (250 mL) water. Store in a spray bottle and shake vigorously before use. To adjust the degree of stiffness, add more cornstarch.
- Bleach alternative: Soak whites overnight in a solution of 1 cup (250 mL) soap flakes, ¼ cup (60 mL) borax and 20 L of water. Or add ½ cup (125 mL) of borax to your wash load.
- Fabric softener: If you wash with soap, add ½ cup (125 mL) to 1 cup (250 mL) white vinegar to the rinse cycle.

Problems:
- Graying may be caused by too little detergent, not enough hot water, an oversized load or insufficient soaking. (See following section: For a Whiter Wash.)
- Mineral spots: hold stained area over pot of boiling water and squeeze lemon juice on it.
- Yellowed clothes can be revived by soaking them in a twelve parts water to one part white vinegar mixture for a few hours.
- Or boil them in hot water with the juice of half a lemon or a few drops of ammonia if stains are stubborn.

- Old yellowed tablecloths can be restored by adding lemon juice with salt added to it for a presoak.
- For badly yellowed clothes only: soak overnight with ½ cup (125 mL) bleach and ½ cup (125 mL) automatic dishwasher detergent to a tub of hot water. For nonbleachables: 1 to 2 tsp (5 to 10 mL) cream of tartar to 1 gallon (4 L) hot water. Then wash clothes as usual.
- Linting may mean overloading or too little detergent.
- To prevent piling, turn garment inside out and use gentle cycle.
- To prevent snags, turn clothing inside out.

To Brighten a Colored Wash:
- Add 1 tsp (5 mL) brown or apple vinegar to wash water to make blacks blacker. This is also good for bone, fawn, brown and navy.
- For blue: add a dash of salt to wash water.
- Cottons: add ¼ cup (60 mL) brown vinegar to final rinse.
- Green: add alum to wash water. Good also for mauve.
- Pink and red: Same as for black, but also add a few drops of red ink to the final rinse.

To Prevent Jeans from Fading:
- When washing, turn the jeans inside out and wash in cold water. Dry on the lowest heat setting.

For a Whiter Wash:
- Soak each load of laundry in cold water for fifteen minutes, then proceed as usual. This really works, but you have to remember to turn the machine back on—mine sometimes sits there for hours.
- For exceptionally dirty kids' clothes, pour liquid detergent on the stain and brush lightly with a scrubber before throwing it into the wash.
- For filthy collars try any of the following before you wash as usual. Rub in a paste of vinegar and baking soda. Or rub with white chalk until the spot turns white. Let stand for an hour and wash as usual. Or brush in biodegradable shampoo; it's designed to dissolve body oils.
- Perspiration stains can be removed by soaking in vinegar.
- Oily clothes, towels or linen: soak in a solution of 1 tsp (5 mL) of lye to 3 gallons (13 L) soapy water. Wash as usual. Caution: lye is caustic.

- Fill the machine with just enough water to cover the clothes. Add ¼ cup (60 mL) bleach, agitate, drain water and wash as usual. Bleach is toxic and you should only use this method when desperate.

Rinsing:
- Add 1 cup (250 mL) vinegar to the final rinse to get rid of lint. This will also dissolve soap and make clothes smell sweet.
- A dash of Epsom salts added to the rinse water will keep colors from fading or running.
- 1 tsp (5 mL) vinegar added to rinse will help panty hose keep its elasticity.
- Remove spots by a mixture of two parts water to one part rubbing alcohol.
- For really dirty work clothes, add ½ cup (125 mL) ammonia to the first rinse water.
- For sweaters, add biodegradable creme rinse to the final rinse water. Or rinse in lukewarm water and add a few drops of glycerine. You can utilize your waterbed to dry sweaters on top.

(*See also* Commercial Cleaning Products; Hand-washing Clothes; Stain Removal: Fabrics.)

COCKROACHES

- A good biodegradable mix is a solution of 1 percent borax to 99 percent water. Mix up in a spray bottle and shake well. Use a coarse spray in all the dark places, cracks, baseboards and anywhere else in the kitchen they might be lurking. Reapply six weeks later. Be very careful when using this and the following mixtures especially if you have children or pets at home.
- A 1 lb (450 g) can of boric acid compound sprinkled in all the usual places should slowly rid a house of cockroaches for up to a year.
- Mix three parts boric acid, three parts sodium fluoride, one part icing sugar and two parts cocoa in a solution and leave wherever cockroaches congregate.
- Paint the inside of cupboards with a solution of 1 tbsp (15 mL) alum dissolved in 2 cups (500 mL) hot water.

(*See also* Insect Control.)

COFFEE MAKERS

You can now buy filters made of unbleached paper and very effective cloth ones that last for a long time. Don't flush coffee grains down the sink—you can block pipes. Instead, put coffee grains in the compost. Reuse paper filters two or three times.

- If your coffee tastes bitter, the machine probably needs cleaning. Mix equal parts of vinegar and water and brew it through the coffee maker, followed by two brewings of cold, clear water. If you do this once a month, you'll eliminate mineral deposits and make better coffee.
- Or run 2 tbsp (30 mL) baking soda and soapy water through a cycle, then do it again with clear water.
- When sediment develops in a glass coffee percolator, put a little vinegar and a few pennies on the bottom. Let sit for a few minutes and shake. Remove vinegar and pennies then rinse thoroughly.

COLDS

Being tired or chilled doesn't necessarily mean you'll get a cold. Old-fashioned cures are still the best:

- Drink lots of fluids; go to bed with a hot water bottle or heating pad, and two aspirins and stay there until morning. However, don't give children or teenagers aspirin, as it has been connected to Reye's Syndrome.
- Make a natural decongestant by crushing four 500 mg tablets of vitamin C. Strain in enough water to fill a dropper bottle.
- Always eat a balanced diet when you try anything extreme.
- For coughs: 1 tbsp (15 mL) honey in a few drops of lemon juice.
- Make a tea of the following herbs and spices: fenugreek, angelica, garlic, horehound and licorice root.

COMMERCIAL CLEANING PRODUCTS

If you prefer the convenience of buying commercial cleaning products rather than making your own, many are available that were developed with the environment in mind. These products can be found in health-food stores, through some large grocery chains and through direct-sales agents and on the Internet.

- Don't purchase any product that doesn't list all of its ingredients. Some detergents, for instance, are petroleum based.

- Avoid aerosols.
- If animal welfare is a concern, check the ingredient list for animal by-products. Soaps frequently contain animal fats.
- If your favorite grocer doesn't stock environmentally safe products, ask for them or write to the head of the store chain.

If you must buy harsher cleaning agents, always read the label. Specific signal words and symbols on the packaging are a guide to how harmful the product is. Anything labeled "Poison" is highly toxic; a "Danger" symbol means the product is extremely flammable, corrosive or toxic; and those with a "Warning" or "Caution" label are less toxic. Always buy the least toxic product for your needs.

Anti-Bacterial Products:

A number of antibacterial products, including dishwashing liquids and hand soap, have been put on the market recently. These products are unnecessary and potentially harmful, as they may cause resistance to common bacteria.

- Use soap and water rather than antibacterial soaps and dishwashing liquids. Soap works by loosening bacteria from surfaces, and running water washes the bacteria away.
- Wash your hands often. Rub hands vigorously with soap, followed by a thorough rinse. Just 15 seconds of handwashing without any special antibacterials will flush 99% of the bacteria off your skin.

COMPANION PLANTING

Use herbs instead of insecticides to get rid of garden pests:

- Ants: tansy and mint
- Aphids: chives and garlic
- Fleas: fennel and tansy
- Fruit flies: basil
- Houseflies: basil, chamomile, mint, rue, tansy
- Mosquitoes: chamomile

COMPOST

The ecologically-sound household will always have a compost pile somewhere, whether it's on the balcony or in the backyard.

- What not to compost: bones and meat—they attract animals and take forever to break down; anything that's been treated with chemicals, and that includes grass clippings that you've poured weed-and-feed mixtures on; any diseased plants or leaves; charcoal briquettes that have been treated to make them light quickly. Use citrus peelings in moderation; the worms who will be turning all this stuff into compost aren't that keen on them.
- To make the easiest compost-maker of all, dig a shallow pit and fill it with alternating green and brown layers; chop things up as finely as you can. In the green layer, put plants, nontreated grass clippings, leaves and weeds (without seeds). In the brown layer, put soil, fertilizer (sheep or cow manure), kitchen leftovers; stuff from the vacuum cleaner. Add a layer of soil or manure to get things heating up. Keep slightly damp and turn in two weeks.
- For faster composting, do the same as above, but build a support around the pile so that air can get at the material. Have an opening at the bottom to provide air circulation and access to the finished compost. You can use a wooden bin with slats 4 ft square (1.2 m^2) or a wire-mesh pen with the same dimensions.
- If compost starts to smell, turn it and add a layer of soil.
- In spring distribute the compost throughout the garden as mulch. It is so benign you can even put it near young plants. (*See also* Mulching.)
- Apartment dwellers can use a process called vermicomposting. Get a small box with holes in the bottom for drainage. Buy earthworms (check the Yellow Pages for suppliers) and make a nest of leaves and shredded paper. Set it on bricks with a pan underneath. Proceed with normal composting. You can leave the bin on the balcony or in a closet. Dig the compost into your plants.

COMPUTERS

Care and Maintenance:

- Be sure the plug you use is not on a circuit servicing heavy equipment, or the one you use for the vacuum, air conditioner, refrigerator or other appliances with on-off cycles.
- Keep covered and dusted. Allow air to flow freely around the machine.
- Never let anything with a magnetic field get near your machine or diskettes.
- Never clean the monitor with glass cleaner while the machine is turned on. You stand a chance of getting a shock.

- Keep the area free of static, but don't spray the machine with antistatic spray. Keeping your house well humidified will not only help your skin, it will also keep your computer static-free.

Environmental and Energy-Saving Tips:

- Turn off your computer when you're not working. If you prefer not to have to turn it off and on over and over again, at least turn off your monitor, printer and lights when you leave the room; or put the computer in "sleep" mode.
- Choose a low-power computer. Check for energy ratings when you shop for a new machine. Consider buying a laptop computer if it's a suitable match to your needs and pocketbook, as they require less power than desktop computers.
- Print less and save paper. You can get more on each printed page by decreasing the type size and leaving narrower margins. Also, printer utilities are available that will print several reduced sheets on one page or print on both sides of the paper. Use a fax modem and e-mail to send documents whenever possible.
- Re-ink printer ribbons and recharge your toner cartridges. Many companies will also let you recycle your cartridges by returning them to the manufacturer.
- Recycle your computer and software. When you upgrade, don't throw old computers and disks away. Donate them. Even a broken computer is useful for its parts or in fix-it classes at technical schools.

Repair Tips:

Many computer malfunctions are caused by simple problems. Before trying to fix problems, however, keep the following in mind:

- You will need to discharge any static electricity buildup you are carrying. Touch the computer case or some other grounded object to do so.
- Turn the power off before you disconnect cables. Unplugging your PC is even better.
- Don't mess around inside the power supply. It can carry a powerful electric charge even if the computer is turned off.

Before calling for service, try the following:

- Make sure everything is plugged in and turned on. It seems simple, but wouldn't you feel silly if the technician pinpointed that as your problem.

- Plug a lamp into your power strip. If the lamp doesn't come on, press the reset button on the power strip and try it again. If the lamp still isn't working, plug it directly into the outlet. If the outlet works, you simply need a new power strip.
- Check all the external connections—both at your computer's ports and at the peripherals (scanner, printer, etc.). If you have spare cables, try replacing faulty ones.
- Read the manual. Even manuals that are dry reading usually have a "Troubleshooting" section. Scan through and try the methods they suggest. If you are wary of suggestions to open the case, call for service.

(*See also* Home Offices; Internet Safety.)

CONCRETE

- Clean a stained area with a strong solution of trisodium phosphate (TSP). Scrub and rinse. Read the directions carefully. Although it is a better option than other chemicals, treat TSP with caution.
- Fix a crack in concrete by first removing any loose material. Wet with a brush and apply portland cement powder mixed with acrylic bonding agents such as Bondex Latex Additive, Fixcrete or Elmer's Glue-All. Wipe off the surplus with a damp cloth before it dries.
- Want to repaint those concrete steps? Remove the old paint with a strong lye solution. Combine 4 tbsp (60 mL) lye dissolved in 1 cup (250 mL) cold water with 2 tbsp (30 mL) cornstarch that's been dissolved in 1 cup (250 mL) water. Brush this paste on the steps and leave for an hour. Scrape off with a piece of wood, a stiff brush and water. Be sure to wear rubber gloves when you are working with substances as caustic as lye.

Concrete Driveways:
To repair small sections of a concrete driveway:

- Clean surface thoroughly, removing lichen, grease and oil. Remove lichen with a stiff wire brush and treat area with lichen inhibitor.
- To cover old concrete, make sure new concrete is at least 2 in (5 cm) thick. Use an epoxy resin to form a bond between the new and old surface. For a new surface the layer you add should be about ½ in (1.25 cm) thick.

- To patch a larger section, buy premixed patching concrete, but be sure to store it in a very dry or hermetically-sealed place.
- For a crack, mix paving concrete using one part thin cement to four parts sand. Wearing goggles, chip out the crack until it is wide. Brush away debris with a wire brush. Paint on bonding agent between cracks. Fill with mortar.

Concrete Floors:

- Maintain concrete floors in basements and garages by regular soap and warm water washing. (*See also* Basements.)
- Keep concrete dust-free with the following solution: 1 cup (250 mL) floor polish dissolved in 4 cups (1 L) hot water. Swish it around and apply with an old rag mop.
- To freshen up concrete, wash with soap and water and rinse clean. When it's still wet, distribute dry portland cement evenly over the surface. Let it dry for an hour and sweep clean. Rinse with clear water.
- If a painted concrete floor is very dirty, let a phosphate-free liquid soap stand overnight on it. Wash thoroughly the next day.
- Unpainted concrete floor: wearing rubber gloves, make a strong solution of 2 lbs (.9 kg) washing soda to 1 gallon (4 L) of water. Scrub and rinse until clean.
- Ink stain: cover the stain with a thick flannel rag soaked with household ammonia.
- Smoke and fire stains: use a pottery or stoneware container for the following mixtures. Scrub area with powdered pumice and water. Dissolve trisodium phosphate (TSP) in 1 gallon (4 L) hot water. Slowly add a paste of 12 oz (340 g) of chloride of lime to the water. Add enough water to make 2 gallons (9 L). Stir and let stand until a clear liquid forms. Skim the clear liquid and soak a thick flannel rag with it. Place it over the stain and cover with glass to slow down evaporation. Repeat.

To Remove Stains from a Garage Floor:

- To remove fresh oil and grease, cover stain with dry portland cement or hydrated lime, then remove. Repeat until stain is drawn out of the concrete.
- Old oil and grease stains should be covered with an absorbing powder (*See* Cleaning Formulas). Moisten with gasoline (it will help the lifting process) and put a piece of canvas over it to slow evaporation. When it's saturated, replace until the stain is removed. Rinse floor thoroughly.

CONDENSATION

Toilet Tanks:

- Paint the inside of the dry tank with a combination of glycerine and alcohol. This will stop condensation for at least six months. A more permanent solution is to glue a liner in the tank. (*See also* Toilets.)

Windows:

- Stop condensation around picture windows or any other fixed windows to help prevent rot. Brush a layer of shellac on top of the sill trim.
- To keep windows free of frost, add ½ cup (125 mL) rubbing alcohol to 1 quart (1 L) of water and apply to outside windows.
- Apply glycerine to a soft cloth and rub over inside of windows.

CONTAINER PLANTING

Container planting is ideal for people who want to garden but have limited space or only a balcony. Check the drainage so you don't make enemies of those below. All of these suggestions work just as well on a patio.

- Troughs planted with a bit of soil bumped up with peat moss make an ideal balcony container.
- Plant with hen-and-chickens (sempervivums and echeverias) or saxifrages. These make the perfect plants for collecting—there are hundreds of varieties and they are easy to grow, needing only regular watering.
- A lovely combination of plants for balconies is the gray and silver of cerastium, or snow-in-summer, and the deep blue of lobelia.
- To make a lighter load for a window box, use individual plants in pots and set them in the box.
- Get sturdy boxes and place your potted plants at different levels so you can replace them easily when they die without upsetting the pattern of your design.
- Be to sure to have a climbing tomato plant. Let it clamber up netting or chicken wire fixed firmly to a railing so tomatoes don't tumble down and injure someone below.
- Annuals such as begonia, nicotiana, browallia and impatiens will do well in shade. Other annuals are sun-loving.
- Parsley and white cascade petunias are a good combination.
- Although they aren't spectacular during the day, mignonette and moonflower have a gorgeous smell at night.

- Tuberose planted in pots with the silvery leaves of Salvia horminum will look good all summer.

COOKING ODORS (See Odors.)

COPPER

- Remove verdigris (the green stuff) by making a paste of powdered chalk (whiting) and methylated spirits. Rub on with a cloth and repeat until the green has disappeared.
- To remove water spots, dip a cut lemon in salt and rub the spots.

(*See also* Pots and Pans.)

CORKSCREWS

- If you find yourself without a corkscrew, hold the top end of the wine bottle under hot water and ease the cork out with a knife.

CORN ON THE COB

- To tell if a cob is fresh, check the part that was attached to the stock. It's stale if the butt is dry or brown. The leaves at the top will be shriveled.
- Never buy husked corn; the flavor will be totally gone.
- To cook, place enough water in a large pot to cover the corn. Add a bit of milk and a pinch of sugar. Bring to a boil and add corn. Let it boil for only three or four minutes. Alternately, bring to a boil, add corn, cover and take off the heat and let sit for eight minutes.
- If you must store corn, put the butt end in water in the refrigerator.

COSMETICS

- Buy products with the least amount of packaging.
- Check labels to see if the product is biodegradable.
- Write to the manufacturers of overly packaged or non-recyclable packaged items.

COUNTERTOP CLEANING

- Add 6 tbsp (90 mL) of vinegar to container of warm water. It will remove sticky residues. Or use the abrasive cleaner suggested in Cleaning Formulas.
- Use baking soda to get scuff marks off counters.

(*See also* Bathrooms: Cleaning and Maintenance; Cleaning Formulas; Commercial Cleaning Products; Kitchens: Cleaning and Maintenance.)

CPR: Cardiopulmonary Resuscitation

The aim of CPR is to restore breathing and heartbeat to a victim—to maintain life until medical aid comes to the rescue. If you have small children, it's wise to take a St. John Ambulance course. In an emergency, do the following:

- Put the child on his or her back and clear mouth of any blood, vomit or foreign objects.
- Gently tilt the head back with one hand and lift the chin up and forward with the other so the tongue doesn't block the throat.
- Pinch nostrils to close nasal passages and seal your mouth over the open mouth (for babies and small children cover mouth and nose).
- Give four quick breaths into the mouth and check that the chest rises and falls.
- Give one breath every three seconds (a little faster for a baby) until the victim breathes again.
- Keep doing this until professional help arrives.
- When breathing returns to normal, turn the head toward you and turn the victim on his or her stomach. Bend the leg nearest to you slightly to help support the body. Never leave the victim alone.

CRACKED PUTTY

- Remove old putty, apply new putty and shellac.

CRAYON MARKS (See Stain Removal: Fabrics.)

CURTAINS

- Take curtains off their runners each spring. Wash in the usual way, but put them in the dryer on the fluff cycle. Hang them up immediately to let all the wrinkles fall out.
- Sheers: most of them can be washed in cold water and then rehung to dry.
- To make a good finish for light curtains, add ½ cup (125 mL) powdered milk to white curtains; or ½ cup (125 mL) bran to the final rinse water for colored curtains.

CUSHIONS AND PILLOWS

- Use old sheets to make new pillowcases.
- By using loose stuffing, you can make a cushion any shape you want. Just make a form leaving part of one side open and fill. Slip stitch to close.
- If any of your cushions have a zippered cover and the zipper gets stuck, draw a graphite (lead) pencil along the zipper and it will unstick.

CUT GLASS

- If white film on the outside of old cut glassware is thick enough to be scratched, dissolve it in vinegar.
- To remove film on the inside of glassware, combine 2 oz (56 mL) of laundry bleach with a bit of sand and swish it around in the glasses.

CUTS

- For a minor cut, wash in clean water, dry the area and cover with sterile dressing or bandage. Allow it to scab over.
- For a major cut, put a thick, clean compress over the whole cut and press firmly. A clean pillowcase or towel will do. Keep steady pressure and don't remove compress. Get help immediately.

CUTTING BOARDS

- Cutting boards or chopping blocks are treasure troves of microbe buildup. A putty knife with a sharp edge makes a good cleaning utensil on butcher blocks. Wet it down and then, using elbow grease, give it a good scraping. Or use a windshield scraper—the kind you use to remove ice. Finish by washing off with a solution of borax and water. Rinse thoroughly.
- Season any board, chopping block or butcher block with vegetable oil. Reapply as required to keep the board from absorbing food odors.
- If the surface gets rough from sharp knives, buy a scraper with a replaceable blade. Scrape it smooth and then brush it again with vegetable oil.

CUTWORMS

- Bacillus thuringiensis dusted around seedlings will help. Look for it at your local garden center or nursery. Or check out labels to see if it's contained in organic pesticides. This is as benign a pesticide as you will find.
- You can also make a trap of cornmeal or bran meal around plants. The cutworm will eat until it swells up and dies.

DARNING

No one seems to darn anymore. It's a shame because repairing clothing is one of the most sensible ways to prolong a garment's life.

- Use a darning egg or an orange to stretch the fabric over.
- Use thread from an inside seam, or match wool as closely as possible. Then work from the right side and anchor thread.
- Reinforce frayed edges with running stitches. Fill the hole with stitching lines running parallel to the fabric. Weave across this at right angles.
- Don't pull either the fabric or the stitches too tight.
  (*See also* Mending Clothes.)

DECALS

Removal:

- To get rid of bathtub decals, soak them in mineral spirits and scrape off with a plastic spatula. Clean as usual.
- Try Avon's Skin-So-Soft.
- Paint with several coats of vinegar, then scrape off.
- To remove decals from bottles or jars, put a little soft margarine around the edges and wait an hour or so. They should come off easily.

(*See also* Furniture Repairs.)

DECKS

- Cedar or California redwood are the best materials for deck construction. They're expensive but naturally resistant to rot and insects. There's more maintenance involved with cedar than with California redwood.

- If you must use pressure-treated wood, be sure to protect yourself when you are treating the ends with preservative—it's toxic.
- You can get rid of dark patches and mildew stains on a cedar deck by using household ammonia and water. Scrub with a stiff brush and hose down.

DEER

If deer are a pest in your garden:

- Hang a bar of deodorant soap in a bag made out of old, not necessarily clean, panty hose.
- Spread blood meal around. Always replace after a rainfall since it's absorbed by the soil. This will also fertilize your garden.

DEHUMIDIFIERS

A dehumidifier should be cleaned at least twice a year.

- First unplug the unit. Remove the protective outer panels to vacuum dust from the inside and scrub mineral deposits away with a stiff wire brush.
- You can use the dehumidifier water in your steam iron.

DIAMONDS: Cleaning (See Jewelry.)

DIAPERS

The most ecologically correct diaper to use is the cloth diaper. Though they are expensive to start with, in the long run they're cheaper and they can be passed on to succeeding generations of children. Many families find that a good compromise is to use cloth diapers on a regular basis and to use disposables only in special situations, like when attending a party or wedding or when the family is on holidays.

To Clean Cloth Diapers:

- Rinse in the toilet and soak in a solution of one part vinegar and ten parts water.
- When you have two dozen diapers in the pail, put the whole lot into the washer. Use a mild pure soap such as Lux or Ivory.
- Keep plastic pants supple by adding a few drops of mineral oil to the rinse water.

Pinning Diapers:

- To make a pin slip through a diaper swiftly, rake it across a bar of soap first. This will keep it from becoming rusty too.

Using Paper Diapers:

- If you must use disposables, don't buy ones made with paper that has been treated with an acid to make them white. Check the labels when you buy them.
- Make sure they aren't perfumed and that they don't have an absorbent substance that turns to gel on contact with liquid.
- Get rid of loose tabs—a baby can stick one of these in her mouth and choke.
- Dump feces in the toilet, never in the garbage—it contains bacteria that will affect the landfill site.
- Never flush disposable diapers down a toilet on a septic tank system.

DIMMER SWITCHES

- Use a dimmer switch to create both a romantic atmosphere and save on electricity.
- If the knob falls off from constant use, apply yellow carpenter's glue to stick it back on.

DISHWASHERS

Never use a dishwasher until it's full. Turn the machine on in off-peak hours. Here are some tips for effective dishwasher use:

- Give dishes a light rinse before putting them in the dishwasher. Most dishwashers have only a limited capacity to catch loose food bits.
- Always use either a phosphate-free liquid or powder detergent made especially for dishwashers—hand dish-washing soap will create a disastrous amount of suds.
- Make sure the water is hot enough before you start the machine—run the nearby water tap until it's hot.
- Save on electricity by skipping the dry cycle. Just turn off the machine and keep the door closed.
- Use short cycles on easy-to-clean loads.
- If a mineral deposit has formed on the inside of the machine, you probably have hard water. Add 1 cup (250 mL) of vinegar to the machine when it's empty. Run through the first cycle. On the second cycle, add 2 cups (500 mL) of vinegar. Then reset

and go through the full wash cycle using your usual detergent. To save energy, put a load of dishes through as well but leave out silver or brass. From now on, add ½ cup (125 mL) of vinegar to the rinse cycle once a week.

- Never put hand-painted dishes or antique glassware through the dishwasher. The latter will develop a film that will have to be eliminated by hand. Heat can also damage delicate glasses and they can be broken by the jets of water.
- The right amount of water is important. There should be more than enough to cover about 1 in (2.5 cm) above the sump area.
- Make sure there is always a little water in the bottom of the dishwasher so the O-rings and seals won't dry out.
- When you go away, pour a thin layer of mineral oil into the base. It will float on top and keep water from evaporating.
- Use the minimum water temperature necessary. This is about 140°F (60°C), though you may have lowered it to save energy or to make it safer for children. To check water temperature, use a meat thermometer under the tap and hold for several minutes.
- Check to make sure the gasket—the material that surrounds the door—is fitted tightly.

(*See also* Commercial Cleaning Products; Energy-Saving Tips)

DISH WASHING

The gentle art of washing dishes by hand has been forgotten in many households. It's very relaxing and, of course, more environmentally sound. Here are some useful tips:

- When washing dishes, start by filling the sink with about 3 inches (7.5 cm) of warm water. Do a few dishes at a time, and as you rinse them, run the water into the soapy dishes, filling up the sink as you go. You'll use less hot water.
- Make your own liquid soap: dissolve one part laundry soap flakes in four parts water. Bring slowly to a boil until soap dissolves. Simmer for 10 minutes. Keep the mixture in a squeeze bottle.
- Add a little vinegar to dishwater; it will help cut the grease.
- Slip good glasses into hot water sideways so they don't expand too quickly and crack.
- To make soapy dishwater last longer, start by washing the glasses first, then the dishes and finally the pots. The idea is to

keep the water as grease-free as possible for the objects that need the greatest cleaning.

- To help prevent breakage of delicate dishes and glasses, place a plastic mat or a towel in the bottom of the sink.
- The vegetable sprayer on the faucet is ideal for a clear water rinse.
- When you wash and rinse tall glasses and wine glasses, place them on the drying counter farthest from your reach so you won't knock them over by mistake.
- Always add one part vinegar to three parts water to rinse crystal, and let air dry.
- Remove stains from china with a quick rub of baking soda before you wash.
- Rinse glasses, dishes and cutlery soon after you've used them.
- Put pots to soak overnight. Fill with mild soapy water if they won't scrub clean easily after a meal.

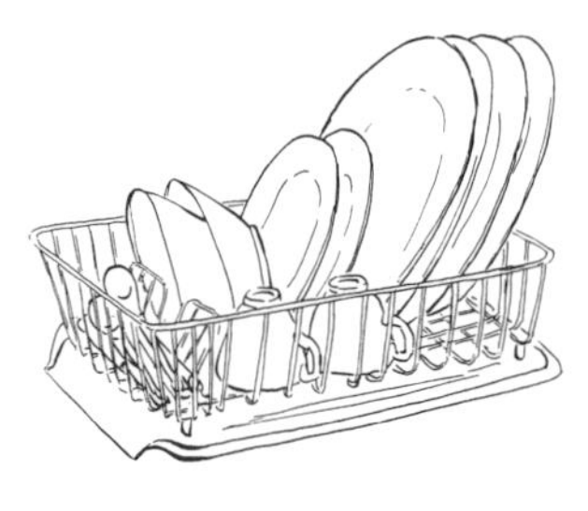

- If you have burnt food in a pot, put soapy water in it, then let it simmer on the stove for a few minutes. This should loosen the food so you can scrub it out.
- Whenever possible, use a plastic scrubber—metal scouring materials eventually break down the protective glaze on most dishes.
- Always use plastic scrubbers and plastic utensils on nonstick pots and frying pans.
- The best drying towels are linen tea towels that have been washed dozens of times to get a soft lint-free surface.
- Polish crystal glasses and silver cutlery, taking care to guard against water spots, but leave the rest of the dishes to air dry.
- Clean cloudy drinking glasses by soaking them for an hour in slightly warm white vinegar. Then use a plastic scrubber to remove the film. If the cloudiness persists the damage is likely tiny scratches (which often occur in the dishwasher). To avoid in the future, hand-wash your best glasses.

(*See also* Commercial Cleaning Products; Pots and Pans.)

DOGS

- Never allow a dog to hang its head out of a car window. Dust and other minute sharp objects will inevitably affect its eyes.
- If an eye keeps getting moist, put a drop of cooking oil in it. If it persists, go to the vet to check it out.
- After you've walked the dog, dispose of animal wastes by flushing them down the toilet rather than adding plastic bags filled with detritus to the garbage.

- Stress and boredom can cause the following problems in dogs: behavioral problems, skin allergies, diarrhea and vomiting.
- When training a puppy, walk it after meals when it's most likely to need a curb.
- Don't get a dog that's too big for your living space. If you have a small apartment, get one that weighs less than 15 lbs (6.75 g) when grown and doesn't need a lot of exercise.
- The following breeds usually don't shed and won't affect those with allergies: Kerry blue terriers, miniature schnauzers and poodles of any size.

Dog Barking:
Nothing is quite as annoying as a dog that barks unnecessarily.

- Be firm. Say "No," jerk the animal's collar and reward with praise when it obeys.
- If your dog barks every time it's left alone, leave a radio on. Or, if it's a puppy, put an old alarm clock wrapped in a towel in its bed—this is supposed to emulate the mother's heartbeat.
- Pretend to leave the house. When your dog starts to bark, go back in and reprimand it. Do this several times a day.
- If your dog barks every time a neighbor goes by, open the door and let it see that this is a friend. Keep reassuring the animal that this is all right. Always reward your dog with hugs when it responds properly.

Dog Fights:
- Distract the animals by throwing a blanket over them, make a sharp noise or hose them down.
- Collaborate with the other owner—each grab one dog's tail and pull.
- When one dog rolls over, this means it's throwing in the towel and giving up.

Feeding Tips:
- Feed puppies three to six months, three times a day; those six to twelve months, twice a day; and adult dogs, once a day.
- Don't overfeed or give dogs table scraps. Always have a bowl of fresh water available.
- Never feed it animal bones, which might splinter; rawhide bones are far superior.

- Approach an injured dog very carefully. Muzzle it with a tie or scarf. If it's bleeding, wrap it in a big towel or old coat to keep it warm and still on the way to the vet.
- Skunk smell *(See* Animal Odors).
- Have your dog checked regularly for heartworms.

DOORS

Most door problems occur when a house shifts, which may cause locks or bolts on doors to move out of alignment with their keeper chambers.

- Check the alignment by marking the end of a latch bolt with a grease pencil. Then close the door to see where the mark shows up on the keeper chamber (also called the strike plate) on the jamb.
- If it's only a bit out of alignment, file away at the keeper chamber and enlarge to make a better fit.
- Make sure there is no dried paint in the keeper chamber.
- If the frame is out of kilter or if the door sticks or binds, try sanding or planing it down a thin layer at a time. Be sure to repaint the exposed wood—especially the top and bottom—or it will warp.
- If the door sticks or binds at the sides, remove the door and slip a wedge of cardboard under the hinge farthest from the spot where the door sticks.
- If you're bothered by a squeaky door, a drop of olive oil or sewing machine oil will do the trick.
- If your patio doors tend to stick when you try to open them, vacuum the tracks regularly and coat with a fine layer of sewing machine oil. Or try using an eraser to clean them up and keep them moving smoothly.
- If a door jams shut, it could mean there is too much humidity in the house. Open a window or turn on the kitchen fan to help dry up the moisture in the air.

(*See also* Frozen Locks.)

Doorknobs:

- If a knob becomes loose on an old doorknob, wrap the spindle in thin, white plumber's tape to give some grip.

Door Insulation: See Drafts

DOWN COMFORTERS AND PILLOWS

Down comforters or duvets are the most useful bedding ever invented. Warm in winter, cool in summer, they also make bed-making easier and are a good investment. When you buy these lovely items for your bed, check the labels to make sure you can wash them. Some are dry-clean only and that warning should be heeded. If you can wash them, try the following:

- Check for any rips in the seams or tears in the fabric shell. Make all necessary repairs before washing.
- Wash them in the delicate cycle and add about half the soap you would normally for a full load. Fill machine with enough water to dissolve the soap completely before you add the comforter.
- Make sure the comforter or pillows are distributed evenly in the tub and add a clean sneaker—it will help balance the load and beat out the clumps of down that inevitably form when the comforter is wet.
- Dry on a low-heat setting and, again, make sure you include the sneaker to help prevent clumps. It makes a terrible noise, but don't be put off by this.

DRAFTS

If you have a nasty draft, try the following to help combat it:

- Fill an old stocking or leg cut from a pair of panty hose with styrofoam packing material. This will keep a dangerous pollutant out of the garbage and do you some good. Stack it against the drafty area.
- Buy a commercial draft evader. They come in many forms, including ones that look like snakes.

Air Conditioners:

- Seal the vent around window air conditioners with caulk or fiberglass sealed with caulk.

Air Leakages in the House:

Sealing up your house will not only make you feel snugger, it will save energy and lower your heating bills. Don't forget, though, that you also need some ventilation for your health and the health of the house.

- To check for possible drafts, use a lit candle and pass it in front of all your plug outlets.
- Do the same with mail slots, windows, patio doors, doors and electric outlets. Any guttering could mean that the caulking has cracked.

Cold Walls:

- A cold outside wall can be warmed up by adding large wall hangings or wall-to-wall heavy curtains.

Doors:

- Leakages around doors can be fixed temporarily by putting weather stripping around the frame until the whole door can be repaired.

Electric Outlets:

- Turn off the electricity, remove the switch plates and fill the hole with some form of fireproof insulation.

Large Windows:

- Make heavy insulated curtain liners that can be removed in summer (unless you need them to keep heat out at that time). Draw them as soon as the sun is off the window to hold in heat. (*See also* Weather Stripping; Windows.)

Mail Slots:

- Hang a piece of heavy cardboard or a double thickness of felt from the top of the slot and let it hang down. This will not obstruct the opening but will hold back the drafts.

Small Windows:

- Seal them with plastic stripping. (*See also* Caulking.)

DRAINS

General Rules for Clogged Drains:

- Pump the plunger half a dozen times before one last plunge and a quick withdrawal. If the water runs down the drain with any speed at all, you've removed the blockage. If not, try again.
- Drain cleaners are almost as caustic as oven cleaners. The best way to avoid using drain cleaners is through prevention. If clogs occur, always try non-chemical treatments first.

To Prevent Clogged Drains:

- Never pour grease or coffee grounds down the drain.
- Prevent clogging in the first place by pouring boiling water down the drain once a week.
- Or pour ½ cup (125 mL) of baking soda followed by ½ cup (125 mL) vinegar with a pinch of salt down the drain. Then rinse with hot water. Do this on a regular basis.
- Pour 3 cups (750 mL) boiling hot vinegar down the drain for a quick clean-out.
- Clean floor drains by removing strainers and reaching down to clear out any detritus by hand—wear rubber gloves if you're squeamish.

Slow Drains:

- Dissolve ½ cup (125 mL) of washing soda (be sure to use rubber gloves) in a pot of hot water. Pour it down the drain and let stand for fifteen minutes then flush out with hot water.
- Or try pouring 1 cup (250 mL) baking soda, 1 cup (250 mL) salt and ½ cup (125 mL) white vinegar down the drain. Cover with a stopper and leave for fifteen minutes. Then remove the stopper and rinse with 1 kettle of boiling water.

Blocked Drains:

- Use a plunger or a plumber's snake followed by a slow-drain treatment. You may need to use the slow-drain treatment more than once, but it will work.

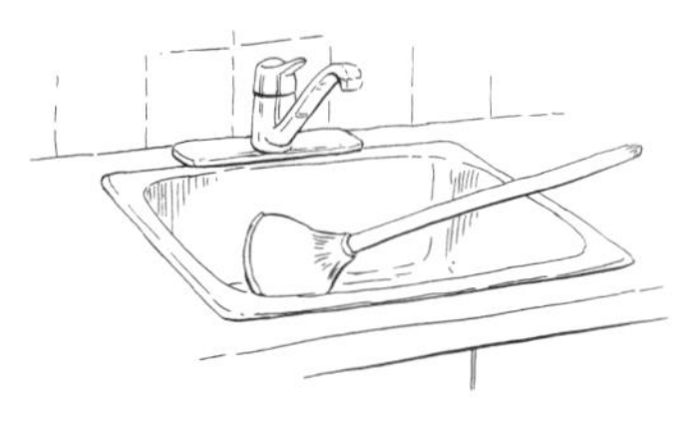

Backed-up Drains:

- Clean out drains with cleaning agents suggested above. Using low-foam types of detergent will avoid this problem.

Bathroom Drains:

- Plug the overflow drain with a wet facecloth. This will help build up pressure if you use a plunger to unclog the drain.
- Thread a thin piece of wire into either the drain or the overflow drain—whichever is closest to the trap—to reach the blockage.
- You can remove a pop-up drain by a quick turn to the left or, often, just by wiggling it.
- To unclog the toilet, bail out the water until the bowl is only half full. Plunge at least ten times. Pour water into the bowl. If the water level rises, bail out the excess and try again.

- If you have a shower stall, keep the strainer clean. If it gets clogged, use a garden hose to release any blockage. Push the hose down the drain, pack cloths around the opening, then turn the water on and off. Don't leave the hose in the pipes—any drop in water pressure will bring the gunk back up to the surface.

Kitchen Drains:

- If there's water in the sink, make sure you have a pot or bucket under the trap to catch it. Poke around the trap with a straightened piece of a clothes hanger and you will probably move the obstruction along.
- Standing water in a dishwasher means a plugged strainer basket or a dirty hose loop (the one that vents the machine). Just clean out the buildup of dirt and grease (*See also* Dishwashers).

Laundry Drains:

- To keep the washing machine drain free, attach a piece of old panty hose with heavy elastic to the machine hose. You'll be astonished at how much stuff this picks up. Clean it or replace it regularly.
- Flush out drain regularly with the cleaners suggested above.

DRY CLEANING

Dry cleaning is a major contributor of chemicals to our ecosystem. When possible, avoid cleaning your clothes this way. And limit your dependence on it by buying clothes that don't need it. As well, many items that say "Dry Clean Only" can be washed with excellent biodegradable products meant for delicates.

If you need a really professional look, get a dry cleaner to do pressing only. However, if you must dry-clean clothing, here are some useful reminders:

- Clean all parts of an outfit at the same time—color changes will at least be uniform.
- Always identify stains to the dry cleaner.
- Be sure to remove the plastic wrap and air clothes out before wearing them so the dry cleaning solvent has a chance to dissipate.
- See if any local dry cleaners use the "Ecoclean" method. Instead of using "perc," the standard dry cleaning solvent,

Ecoclean is a combination of spot cleaning with environmentally friendly soaps, tumbling, steaming and pressing.

- Be aware that sequins can melt in the process. Remove trim before cleaning.
- Leather has to be cleaned by a special process. See Leather for home cleaning methods. Don't store leather in plastic—it won't breathe properly.

DRYER LINT

Don't just throw out dryer lint. It has many uses:

- Lint stuffed into old pantyhose is effective against drafts.
- It is a good stuffing material for homemade dolls and toys (unless children are sensitive to dust).
- It makes for great clouds on school art projects. You can get really funky colors, too, like grey for storm clouds and reds for sunsets.
- Shoe stuffing: Excellent for sore toes or if your shoes are a little too loose.
- Nest material for birds: Put out near your birdfeeder in the early spring.

DRYERS

- If the dryer works but the clothes aren't drying properly, check the lint trap near the front door and the duct at the back to be sure they are unobstructed. Check the outside vents for any obstructions. This should be done regularly anyway. Check fuses (one controls heat and one controls the drum).
- Dryer Vent Hood: This apparatus must swing open easily with the force of air. A drop of sewing machine oil will help keep it movable.

(*See also* Clothes Drying.)

DRYING FOODS

You can dry most foods successfully. Prepare them as you would normally—peeling apples, for example—and then slice.

- Tomatoes don't need peeling, just slicing.
- Dip peaches, cherries, nectarines, apples, pears and bananas in lemon juice to preserve color. Or dip in 1 quart (1 L) container of water with 1 tsp (5 mL) ascorbic acid added.

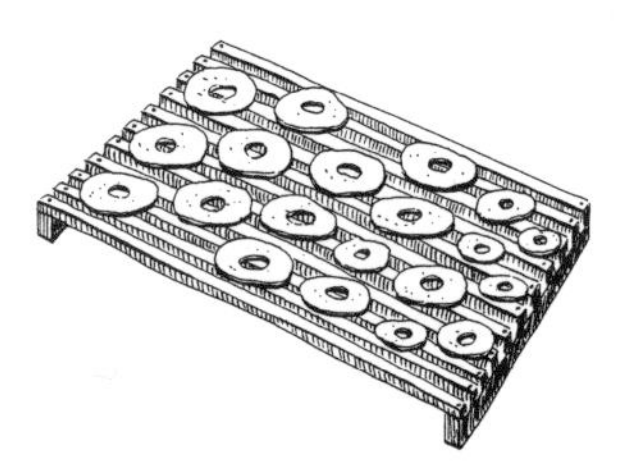

- Blanch fruits and vegetables before drying.
- In summer, put fruit or vegetables on trays of wooden slats and set off the ground in a sunny spot. Turn once a day. Cover trays at night and protect them from animals. It will take three to five days for the food to dry.
- In bad weather, use a commercial food dryer or the oven kept at 95°F (35°C) to 145°F (63°C), allowing three to four in (7.5 cm to 10 cm) between racks. Turn for even drying.
- Leave dried food covered for five to ten days. If condensation appears, dry a few hours more.

DRY ROT (See Wood Rot.)

DRYWALL (See Plasterboard or Drywall.)

DRY WELLS

A dry well will allow extra water to seep away from downspouts or patios to keep it from entering foundations. It will also keep a low-lying garden drier.

- Make a dry well out of concrete blocks or bricks. Dig a square hole about 5 ft (1.5 m) deep and 2 ft (.6 m) wide. You can fill it with rubble or line the edges in a honeycomb pattern with bricks or concrete. I made an attractive top to my dry well by laying boards across the top in such a way that it looks like a little bridge. At any rate, make sure you have a safe top so no one will fall in.
- Or remove the top and bottom of a barrel and lower it into a hole filled with rocks. Cover with a sturdy top for protection.
- Connect with a drainage pipe to the area you want drained.

DUSTING

- Dust before you vacuum.
- Use a clean, dry cloth that won't scratch—old tea towels or cotton t-shirts are soft and lint-free.
- Work from the top down so you don't just redistribute dust.
- Soak your dust rag in a mixture of 1 tbsp (15 mL) glycerine and 1 tbsp (15 mL) vinegar for an hour. This is also a good way to keep mirrors from misting up.
- Cover the head of a mop with a dust cloth to remove cobwebs in corners.

(*See also* Furniture Care and Maintenance.)

EARTHQUAKES

Do the following during an earthquake:

- Keep away from windows, mirrors and ceiling fixtures.
- Stand in a central hallway or get under a desk, table or bed.
- Don't go outside, but if you're already there, keep away from power poles. Get to an open area.
- If you're in a car, pull over immediately.

Once the Quake is Over:

- Turn on a battery-driven radio for instructions.
- If you smell gas, turn it off at the main. Open windows and evacuate the house.
- Don't flush toilets until you know sewage lines are clear.
- Boil nonbottled water for twenty minutes.

EARWIGS

There aren't very many bugs that irritate me, with the exception of earwigs. They can bite.

- If they should come in from the garden, crush bay leaves and put them along baseboards and windowsills.
- To make a trap, leave a piece of rolled up newspaper overnight and then drop the offending beasts that shelter under it into a pail of soapy water.
- If they get into the house, control them with one of the insecticides listed under Insect Control.

EAVESTROUGHS

Well-maintained eavestroughs can cut down on the problem of dampness in the basement.

- Clean eavestroughs after all leaves have fallen. It's a mucky job and requires getting up on a ladder. Scrape them out with a putty knife or trowel. Install screening to keep leaves out next year!
- Make sure downspouts are free of blockage and install a leaf strainer at the top. If you find a blockage, free it with a plumber's snake.
- Flood the gutter with water from the hose to make sure it's running free.
- Make sure water runs away from the foundation.

EGGS

- Two yolks will substitute for one whole egg. In cake batters, just add a bit more milk. Replace one in three eggs with 1 tbsp (15 mL) cornstarch if you are short an egg.
- A cracked egg can be boiled by wrapping it in aluminum foil then boiling as usual. After cooking remove foil immediately.
- If an egg cracks while cooking, put 1 tsp (5 mL) salt, lemon juice or vinegar in the water.
- Never wash eggs before storing them. The fresher they are, the duller they look.
- To test the freshness of an egg, put it in a pan of cold water. If it lies on the bottom, it's fresh. One end slightly raised means it's moderately fresh. If it floats, it's probably too stale to eat.
- If you open an egg and it has a very watery white, it's not fresh.
- Bring eggs to room temperature before using them; the whites won't be tough. Or put in a pot of warm water for ten minutes.
- The best scrambled eggs are made with only 1 tbsp (15 mL) of cold water added and cooked over the lowest heat possible in the top of a double boiler that's been rinsed with cold water. It takes ages but is worth it.
- Perfect hard-cooked eggs: use large eggs (any number) and cover with cold water. Bring to a boil, turn off the heat and let sit for six minutes. Cool by putting into a bowl of ice water. The yolks will still be moist. If you like them harder, add another minute.
- To shell hard-cooked eggs easily, bang on the side of the sink while they are still hot until the whole surface is a maze of tiny cracks.

- If you drop an egg, cover it with salt and leave for fifteen minutes. This keeps it from running all over the place and makes it a snap to pick up.
- If an egg white refuses to whip, add a touch of glycerine or a pinch of baking soda.
- Any left over whites or yolks can be frozen in ice cube trays. Or cover extra yolks in milk and they'll keep in the fridge for three days.
- If you have lots of egg whites, mix them up for a refreshing face mask. Leave it on until it's a slick surface and then rinse off.

ELECTRIC CONSERVATION (See Energy-Saving Tips.)

ELECTRIC HEATERS

- Check the efficiency of your electric baseboard heater by placing a block of wood against it to cut off the airflow. If the heater doesn't shut off within a few minutes, something is wrong, especially if it scorches the wood. There might be a faulty plug here and it should be checked out and replaced. Another possibility is that the thermostat is malfunctioning.

ELECTRICITY—UNDERSTANDING THE TERMS

- Ampere: Amp for short is the measurement of the amount of electricity flowing through the wires. Fuses, outlets and switches are rated in amperes.
- Volt: A unit of electric pressure. It forces the amperes of current to flow through wires.
- Watt: The measurement of the electrical power used.
- Amps × Volts = Watts. Electricity is measured in kilowatts or units of 1,000 watts. One kilowatt hour equals watts consumed over one hour. Meters register the amount of power consumed in kilowatt hours.

ELECTRIC SHAVERS

- Keep it clean and in good condition. Always blow out stubble after each use.
- If your shaver stops working, check the head. Look for defective blades or gaps in protective screens. Replace blades or blades and screen as a unit.

ELECTRIC SHOCK

- Do not touch anyone who has received an electric shock from an appliance or live wire—it could be transmitted to you. Shut off electricity immediately or pull the plug. Use anything that will break the contact but won't conduct electricity—a wooden broom or a chair will do.
- If there is any water around, don't step in it.
- Check to make sure the person is breathing. If breathing, put the patient in recovery position: face to one side, on stomach with one leg slightly raised on a pillow to support the body. Otherwise treat for shock or rush to a hospital immediately. (*See also* shock.)

ENERGY-SAVING TIPS

The credo of the environmental movement is Reduce, Reuse and Recycle. When it comes to energy, it's vital that we all reduce our energy intake.

- It only takes 1/20,000 of a second to turn a bulb on. Not even a blip on your hydro meter. Always turn off lights when not in use.
- Only use heavy-duty electricity guzzlers such as dryers and dishwashers at off-peak hours. In some areas, rates are lower at these times.
- Turn off the dryer cycle on your dishwasher and let the dishes dry with the door closed.
- If you have clean reflectors under burners, they will radiate back more heat. Keeps burners themselves cleaner as well.
- Don't bother preheating an oven unless you are baking something that requires precise timing.
- Never place a fridge near a source of heat and keep the temperatures as low as possible. Turn off the butter conditioner—it's only a little light. Thaw frozen food in the refrigerator to help keep the fridge cool.
- Switch to fluorescent lights or buy expensive energy-saving bulbs—they pay off in the long run.
- Use dimmer switches on lights.
- Front-loading washing machines are the most energy-efficient washers.
- Always use a cold water rinse. In fact only use cold water for laundry when it's possible (dark and slightly soiled laundry is fine in cold-water washes use warm-water wash for whites). You'll save money, too.

- Don't overuse the dryer. Experiment with the length of time needed for various fabrics. Most clothes will hang out better with fewer wrinkles if they're a bit damp. If you're buying a new dryer, look for one with a humidity sensor—it will turn off when the clothes are dry, saving you money on your electricity bills.
- Wall-to-wall carpeting keeps a room warmer, quieter and dryer.
- Keep your thermostat at 68°F (20°C) or lower. At night and when you leave the house for a whole day, turn it down to 60°F (16°C).
- A duvet is more energy efficient than an electric blanket.

(*See also* Air Conditioners; Clothes Drying; Computers; Dishwashers; Heat Conservation; Home Offices; Lighting; Refrigerators; Stoves and Ovens.)

FANS

- To find out if you have a bent blade, insert a pencil through the grill and hold it so that the leading edge of one blade just touches the pencil's point. Turn blade by hand—a bent blade won't touch the pencil. (*See also* Ceiling Fans.)

FASTENERS ON CLOTHES

- Sew on the snap point first. Take a piece of chalk and touch the point. Turn the material over and rub with a finger to mark the spot where the snap should be sewed on.

FENCE MAINTENANCE

- Pressure-treated redwood, cypress and cedar will age beautifully when exposed to weather. Cedar has the added advantage of being resistant to rot and assault by insects. Applying a coat of transparent stain will protect against fading caused by ultraviolet rays.
- Regularly treat fences with organic preservatives to hold off rot and animal infestations and to preserve natural wood color. Aura, a German product, is excellent if the wood hasn't been satined. It's available at environmental stores.
- Soak fence posts with creosote for two days if you aren't sure they've been treated. However, don't put them near the house or delicate plants; this stuff is poisonous. Avoid getting it on your skin—wear gloves. If you have a garden you value, try to find organic preservatives. Any petroleum oil is highly toxic and not good for the soil. Again be careful about reading contents.
- Create a bath for the wood by making a square of old bricks and lining it with heavy plastic that doesn't have any tears or holes in

it. Place the wood in the bath, pour preservative on it and leave for a minimum of two days or up to a week if you have the time.
- Painting the fence with preservative as you build it will save time and you won't miss any spots.
- Make sure you put preservative on the end grains.
- Weatherproof tops of fence posts by capping them with hardwood or plastic and topping with a decorative finial.
- Fence posts generally rot first. Add a reinforcement post next to the rotten one and cut off the old one where it's rotten. Remove the old post. Pour concrete or drive in metal footings for the new post.
- Always use a good quality outdoor paint. Use a base coat and two top coats.

FIREPLACES AND HEARTHS

A fireplace is a luxury in a time of waning supplies of wood and increasing air pollution in larger cities. Light a fire only on special occasions or when no other heat source is available, and be sure to keep the fireplace clean and running efficiently.

- If you have a cement brick fireplace, you can paint it with exterior latex paint. It can be brushed on easily.
- If your house smells of smoke whenever you light a fire, your house may be too tightly sealed. Just open a window.
- Put salt on logs and you'll reduce the amount of soot produced.

Cleaning:
- Sponge quarry tiles around a fireplace with vinegar and they'll come up shiny. Don't use vinegar on marble.
- If the rock facing of your fireplace has smoked areas, rub them with an art-gum eraser.
- Clean a solid brass firescreen with steel wool and clean the same as for copper. (*See also* Copper.) Most firescreens today are brass-plated, though, so don't use this method on them—you'll strip them if you do.
- Scrub the hearth with a stiff brush or one of the metal brushes used on bricks dipped in a solution of 1 tbsp (15 mL) of trisodium phosphate (follow the directions closely) in 1 quart (1 L) of warm water.
- To keep fire irons from rusting, make a strong paste of fresh lime (it's caustic, so wear gloves) and water and apply thickly. This will help preserve them if you're away for a long spell.

Energy-Saving Tips:

- The flue on a fireplace should be one-tenth of the whole opening for greatest efficiency.
- If you have a thermostat in the same room as a fireplace, turn it up a few degrees whenever you use the fireplace and you won't freeze out the rest of the house. Better still, move the thermostat to a more suitable location so you can turn it down whenever the fireplace is on.
- Don't burn loose newspapers in the fireplace. They create too much smoke and burn too hot, creating a lot of creosote. Instead, make tightly bound logs out of them to use as kindling.
- Make small fires rather than great roaring ones.
- Always leave the flue closed when you aren't using the fireplace and stuff any gaps with insulation. More heat goes up a chimney than you'd imagine.

Fireplace Doors:

- To cut down on heat loss from the fireplace, put glass doors on the front. Glass doors can heat up to very high temperatures, however, so keep children and pets away from them to avoid burns.
- A cloth moistened with vinegar will clean off glass fireplace doors with ease.

FIRE SAFETY TIPS

- Curtains and towels blowing against a heater can start a fire.
- Use night-lights in children's and older people's bedrooms.
- Tape the connection between cords and extension cords.
- Use electric outlet covers on all unused outlets.
- Never overload extenders in plugs.
- Never substitute a different wattage of fuse when replacing.
- Never use frayed electrical or extension cords.
- Keep any combustible material away from water heaters, fireplaces, furnaces and dryers. These materials could include newspapers, household chemicals, rags, curtains, sofas, chairs and rugs.
- Beware of dangerous heaters such as electric space heaters and kerosene heaters.
- Make sure you have a solid and secure fire screen on the fireplace.
- Ensure there's a spark arrester on the chimney cap of a working chimney.
- Have your chimney cleaned and inspected regularly.
- Install smoke detectors and be sure to test them on a regular basis.
- Have a fire extinguisher on each floor.

FIRE SURVIVAL

- Have escape routes worked out on each floor.
- If there is a small fire, try to fight it. For example, if something on the stove catches fire, throw salt, baking soda or baking powder on the flames. Have a large shaker filled with any of these ingredients sitting on the stove at all times for this purpose.
- If a fire gets out of control, leave, close all doors behind you and don't try to take valuables with you.
- If your clothes catch on fire, don't run. Wrap yourself in a coat or blanket, or roll on the ground to smother the flames.
- If a door feels hot or you see smoke, don't open it. Stuff bedding at the bottom of the door and try to get out by a window.

FIRST AID

Keep a box containing the following items in a safe dry place:

- absorbent cotton
- box of assorted sized bandages
- sterile gauze
- large piece of clean white linen (for burns or scalds)—an old pillowcase will do.
- scissors
- safety pins
- tweezers
- alcohol for cleaning wounds
- calamine lotion for sunburns and stings
- mild antiseptic cream
- thermometer
- eye ointment

Make periodic checks of the kit's contents. Make sure there is an ample supply and that creams and ointments haven't passed their expiration dates. Take the kit with you on holidays and even on short driving trips.

Tips:

- To remove a splinter: put an ice cube on the splinter. It will numb the area and allow the splinter to be removed easily. Squeeze a little blood to cleanse the wound.
- To take off adhesive tape, saturate it with baby oil and it will come off painlessly.

- If you pull a muscle, apply a bag of frozen peas to the sore area—it will conform to the shape of your body and won't drip all over you.

(*See also* Bruises; Burns; Choking; CPR: Cardiopulmonary Resuscitation; Cuts; Electric Shock.)

FISH

Tips on Buying and Storing Fish:

- Whole fish should have red gills, clear eyes and moist scales with a firm flesh.
- Fillets should never smell fishy.
- To store fish, wash in cold water, wrap in plastic and put on ice in the fridge.

Tips on Cooking Fish:

- Never-fail fish: turn oven on to 400°F (200°C); drizzle the fish inside and out with extra-virgin olive oil, a few grindings of black pepper and a squeeze of lemon. Oil the pan and place fish in the hot oven. Cook ten minutes for every inch (2.5 cm) on the thickest spot on the fish. It will be perfect and not dried out or crumbly.
- To barbecue fish, place on a very hot grill and cover. Cook three to four minutes, then reposition at a 45° angle. Cover again and cook another three to four minutes.
- Thaw frozen fish in milk to take out the chemical frozen taste. Or soak it in vinegar and water to recover the original sweet taste.
- Wash your hands in vinegar and water to get rid of any fishy smell. Or rub them with lemon slices.

(*See also* Shellfish.)

FISH: As Pets

Fish are ideal pets for an apartment dweller as long as they have been bred for this purpose and not taken from the wild.

- Be sure to get a tank large enough to provide an interesting environment; add shells, places for the fish to swim among and to rest quietly in the shade.
- A good aerator is a must for an aquarium. If you have a fish-bowl, be sure to change the water once a week.

FLAGSTONES

- Cracks in flagstones mean it's time to replace them. It's easier to do this than to try and repair them.
- For longer life, get concrete cast as flagstone.
- Instead of using mortar, plant ground covers in the joints if you have an earth base or sandy-earth joints.

FLEAS

Chemical-based flea collars for dogs and cats are harmful. There are herbal collars that have some effectiveness and they are much safer. Vigilance is the best method. Try some of the following:

- Wash your cat or dog with a nontoxic mixture of orange oil and shampoo to remove the fleas. Repeat weekly until the infestation is over.
- To relieve your pet's itching and help heal flea bites, you can try this herbal rinse. Mix 1 tbsp (15 mL) each of dried chamomile flowers, dried peppermint and dried calendula flowers with one quart (1 L) boiling water. Allow it to stand until lukewarm, then pour it liberally over your pet. Make sure the animal's coat is soaked through to the skin. Then allow your pet to drip dry.
- Also wash your pet's betting often.
- Vacuum your house and pet very carefully and then get rid of the vacuum bag immediately. Eighty percent of the fleas will not be on the pet.
- Liberally sprinkle salt on carpets and areas where your pet wanders. Leave for ten minutes and vacuum.
- Spread diatomaceous earth around baseboards and carpets, and rub your pet with it to dry up fleas. It is available in bulk from environmental stores.
- Add 1 tsp (5 mL) powdered brewer's yeast (that's brewer's yeast, not baking yeast!) to your pet's food to keep fleas from developing. Fennel, rue or rosemary added to your pet's food will also help eliminate fleas.
- Although it's not scientifically proven, some people swear that adding apple cider vinegar to your pet's water dish is an effective treatment, too.

FLIES

- The safest way to get rid of flies is to use an old-fashioned fly swatter.
- Don't leave any food around for flies to feed on.

- Make sure the kitty litter is clean—flies love it.
- Sprinkle borax around the inside of garbage cans.
- Flies will be repelled by a sponge dipped in hot water and dashed with oil of lavender. Leave it in the middle of a room and rewet a couple of times a day.
- Cloves heated in the oven and left in a saucer by a window will discourage flies from coming in.
- Rub the framework around windows with household ammonia.
- Keep fresh mint in a vase near a window.
- Make your own flypaper strip: dip yellow paper (which attracts them) in honey and hang in the middle of the ceiling.

FLOORS

Brick:

- To repair a brick floor, use a paintbrush to put a film of water into the cleaned-out joints.
- Apply the mixed mortar by using one trowel as a serving dish and another to scoop out a portion of mortar.
- If you slop any mortar on the bricks, sponge it off right away with clear water.
- If the mortar leaves a shadow on the bricks afterward, scrub with twenty parts water and one part muriatic acid and rinse well with clear water. When you are using something caustic such as acid, be sure to wear rubber gloves, and have good ventilation so you don't breathe in fumes.

Slate:

If you have beautiful slate floors or slate slabs outside, keep them fresh this way:

- Use phosphate-free liquid detergent and scrub the slate down to remove any dirt. Rub with boiled linseed oil and allow it to soak in for twenty minutes. Wipe dry. Leave it for twenty-four hours, then reapply oil and wipe dry again. You may need to brush it to get rid of whiting.
- Use half a lemon dipped in salt and rub on stain. Let sit for a few hours, then wash with clear water. (Do a test on a hidden part first.)

Vinyl:

- Laying sheet vinyl: leave vinyl in room for forty-eight hours before laying to let it loosen. Fit the longest wall first.

- If vinyl comes loose, work special cement (available at hardware stores) under the loose edge. Stick some heavy books on top and let it dry for a day.
- To lay tiles, use a rolling pin. The weight will be distributed evenly over the tile.
- Use a hair dryer to soften glue if some slops over and dries.
- To seal seams, run a strip of sticky tape along it and paint shellac over it.

Hardwood Care and Cleaning:

- Polished floors can be cleaned with a mop dampened with vinegar and water (wring the mop out well).
- Don't use water on hardwood floors—ever. Use an environment-friendly cleaner specially made for hardwood floors.
- To dust hardwood, put an old piece of panty hose over the mop and dust. You can wash and reuse when you're finished and the mop remains clean.
- Walk around in slippers with chamois-like soles to buff waxed hardwood floors. Don't do this on urethaned floors—you'll just grind the dirt in. Give the slippers to the kids or guests if you have the moxie.

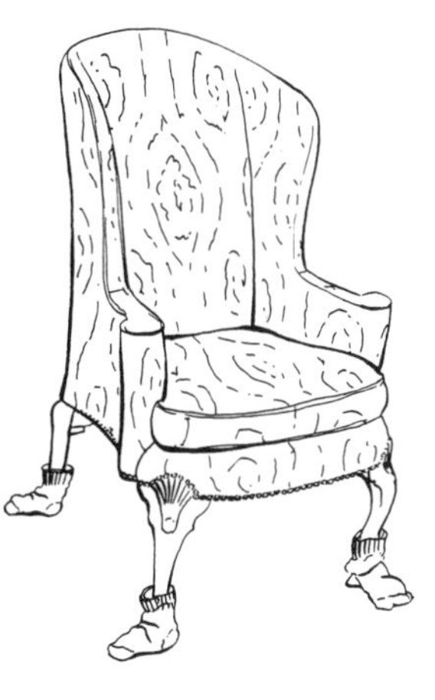

- Remove heel marks with an ordinary eraser. It is surprisingly effective.
- Eliminate squeaky floors by dusting talcum powder into cracks. Or run glue into cracks.
- For old hardwood: sand with a fine-grade paper. Stain and then apply at least two coats of urethane varnish. Dry for approximately twenty-four hours (humidity and temperature will be factors), buff and then apply the next coat.
- Wood floors with a urethane finish may be brought back to life with a light sanding and a fresh coat of urethane.
- To touch up worn areas, roughen the surface with #00 or #000 steel wool and apply coat of urethane. Dull overly glossy areas with #0000 steel wool.
- Fading floor? Put a little brown shoe polish in with the floor wax. Dab it on the dull spots and buff up.

Protect Floors:

- Protect floors by putting socks on furniture legs when you move them around.
- Bunion pads on the bottom of chair legs will act as a protector for your floors.

- If metal protectors on furniture become rusty, sand the rust away. Then coat with clear nail polish.

Repairs:

- Use small brush strokes following the grain of the finish whenever you work on hardwood.
- For a minor scratch, use a small brush and no more than a thimbleful of urethane to paint over the scratch.
- If the scratch is deeper, sand lightly with steel wool and mineral spirits, followed by a rewaxing of the damaged area.
- For a deep scratch, remove wax and fill the scratch with wood putty that is color-matched to the floor. Let dry then sand, vacuum and apply a coat of wax or varnish to match the existing finish.
- Try wax sticks in several colors to get the same finish. Cut out loose strands of wood, heat the wax and drop into the scratch. Spread. Then rub with an ice cube to harden, remove excess with a razor blade. Repeat and buff with a soft cloth. It will be invisible.

FLOWER ARRANGING

- Cut flowers early in the morning using a sharp knife. Immerse stems immediately and let sit in deep water for a few hours.
- To arrange, clip off bottoms one at a time—on an angle and then split stem for more water absorption. Do this under water.
- Strip stems of leaves that would be under water.
- Hair rollers placed at the bottom of a vase will act as a holder for an arrangement.
- Hold very long-stemmed flowers erect by crisscrossing tape across the mouth of the vase.

FLOWER CARE

- To revive flowers, plunge stems into boiling water; by the time the water is cold the flowers will have revived. Then cut the ends again and put them in fresh cold water.
- To extend the life of a bouquet keep it in a solution of 1 tbsp (15 mL) vinegar and 1 tbsp (15 mL) sugar with 1 quart (1 L) of water. The vinegar slows deterioration and the sugar feeds the flowers.
- Carnations like a little boric acid.
- Anemones, lilies and gladioli love a bit of vinegar in the water.
- Geraniums like coffee grounds (after they've been rinsed).
- A penny dropped in a vase of tulips will keep them from opening too quickly.

- Asters, cosmos, delphiniums, peonies and sweet peas like a bit of sugar added to water.
- Add salt to the water of begonias, marguerites, roses, snapdragons, stocks and violets.

FLOWER DRYING

- Mix ten parts cornmeal and three parts borax and bury flowers in this for two weeks.
- Pound the stems of branches to create a greater surface and stand in a solution of two parts water and one part glycerine for a week. Leaves will last for ages.
- Use the microwave oven for drying. Put flowers in a microwave-safe casserole dish with silica gel on the bottom and sprinkled on top. Put 1 cup (250 mL) of water in the back of the microwave oven. Then heat for thirty to sixty seconds.

FLUORESCENT LIGHTS

- A new fluorescent tube may flicker for an hour when first installed.
- Fluorescent lights have a longer life if they are not turned off for three hours after they've been turned on. You're not saving energy by turning them off every time you leave the room.
- Most fluorescent lights won't work efficiently in a room below 50°F (10°C).

(*See also* Lighting.)

FOOD SAFETY

Here are some signs you should look for to make sure what you eat is safe:

- Beef that turns brown in the refrigerator after a day or so is safe but don't keep it for longer than two days, then cook well. If it smells bad, ditch it.
- Fruit that looks bruised or moldy is still safe, just cut the offensive parts away.
- Ham that's been thawed outside the fridge may be dangerous. Neither the smoking nor salting protects it against bacteria. All meats should be thawed in the fridge.
- If the vacuum hasn't been broken on the package, hot dogs are safe for weeks, but once the seal is broken, eat within one week.

- Ketchup or mustard with a black ring around the top of the jar is safe; just wipe it off. It's only a chemical reaction to air.
- A jar of mayonnaise that has been opened in the fridge for six months is still safe as long as it hasn't been left out for long.
- Nuts that look or smell moldy produce aflatoxins which are cancer-causing agents.
- A baked potato left out overnight is not safe to eat even if you reheat it—botulism might still lurk in there.
- Green-skinned potatoes are still all right, but you have to cut all the green away. It contains solanine and can cause gastro-intestinal illness.
- Potato eyes should be cut out because they are poisonous.
- Poultry juices of any kind on any surface are dangerous. Don't touch uncooked chicken and then your mouth. You could get very sick. There may be salmonella in fowl.
- Shrimp that's been cooked but not deveined is okay, but never eat them raw in this condition. The vein is the intestine and it contains waste products.
- Steaks thawed in the fridge and then refrozen are still safe, but thaw in the fridge a second time and cook well. It's better to cook and then refreeze if you must.
- Be sure to wash all fruits and vegetables to get rid of any pesticide residues. Look for a supplier who farms organically. You will find that the produce tastes much better even if it isn't always as perfect as those grown with chemicals. (*See also* Pesticides: On Produce.)

FOOD STORAGE

- Avoid storing food in anything containing lead. This includes glazed pottery dishes or any object that smacks of the happy hobbyist.
- The only embossed glasses you should drink from are those with silver or gold.
- Use heavy plastic storage containers or reuse plastic food containers rather than foil, plastic wrap and plastic bags.
- Food to be fried or sautéed should be taken out of the refrigerator and brought to room temperature first.
- Green vegetables: shake off excess water and store in the crisper. Eat within six days.
- Root vegetables will keep for two weeks in the refrigerator.
- Rutabagas, squash, onions and potatoes: store in cool dark places.

- Citrus stored at 60°F (15°C) to 70°F (21°C) in refrigerator will keep three to four weeks.
- Berries and tomatoes will keep three days if uncovered in fridge.
- Other ripe fruit can be kept at room temperature for three to five days.
- Bread with no preservatives can be kept in refrigerator up to five days. Bread with preservatives can be kept five days in the original bag at room temperature.
- Keep all dry staples as well as oil in airtight containers away from heat and light.
- Oil: keep large containers in refrigerator. It will cloud up but clear again when at room temperature.
- Meat: loosen wrap it comes in or store in wax paper in coldest part of refrigerator. Use within three to four days.
- Any ground or organ meat should be used immediately.
- Eggs, milk and cheese: use within one week.
- Wrap all moist foods and liquid stored in refrigerator.
- Store root vegetables in a picnic cooler in the basement or any area where it's a few degrees above freezing, which is as good as a root cellar and the vegetables last as long.

FOOD SUBSTITUTES

- Make all-purpose flour by using 1 cup (250 mL) plus 2 tbsp (30 mL) cake flour.
- Make cake flour with 1 cup (250 mL) less 2 tbsp (30 mL) all-purpose flour.
- Make potato flour: add ½ cup (125 mL) instant mashed potato flakes to every 4 cups (1 L) of flour.
- Make your own whole wheat flour: 5¼ cup (1,300 mL) all-purpose flour, ¾ cups (180 mL) wheat germ, and 1½ cup (360 mL) natural bran.
- Make butter by beating 2 cups (500 mL) evaporated milk.
- Make buttermilk or sour milk by adding 2 tsp (10 mL) lemon juice or white vinegar to 1 cup (250 mL) of milk. A slightly more complicated mixture but still effective: 1 tsp (5 mL) lemon juice and ½ tsp (2.5 mL) baking soda to 1 cup (250 mL) of milk.
- Make baking powder: substitute 2 tsp (10 mL) cream of tartar, 1 tsp (5 mL) baking soda and ½ tsp (2.5 mL) salt for each cup of flour.
- Make confectioners' sugar by buzzing ordinary sugar in the food processor for a few seconds.

- If you are out of wild mushrooms, cover the bottom of a casserole with clean young grapevine leaves (you can probably find them in your garden or buy them by the jar). Dab ordinary mushrooms with butter, top with leaves, cover and bake in a moderate oven for twenty minutes.
- In pie crusts, you can substitute peanut butter for shortening in pecan, butterscotch pies or butter tarts.
- Make sweetened condensed milk by mixing ¼ cup (60 mL) hot water with ¾ cup (180 mL) granulated sugar. Slowly add 1¼ cup (300 mL) dry skim milk powder. Refrigerate twenty-four hours.

(*See also* Eggs.)

FOOD THAWING (See Thawing Food.)

FOUNDATIONS

Foundations should be a minimum of 8 in (20 cm) above soil level.

Crawl Spaces:

Make sure they are kept well ventilated.

Footings

A footing is a foundation constructed below ground.

- Find out what your frost-line depth is before creating a new footing and add this to the footing depth.
- You can use earth as a footing form if it's firm and clear of rock.
- Allow a curing process of no less than four days.

Maintenance:

- Make sure gutters and downspout work properly and check after hard rain. (*See also* Gutters.)
- Do not install plants closer than 2 ft (.6 m) from the foundations—roots retain moisture.
- Make sure soil slopes away from the house.
- Check for moisture stains, insects, rust, peeling or cracked paint.
- Check bottom of wall in a finished basement for mold, warping and mildew.

- Check under window air conditioners to make sure moisture isn't entering the house from there.
- To test for condensation: tape a mirror to the foundation wall on the inside. Seal edges with tape. If there is condensation on the mirror, you've got trouble. If the surface is dry, but the back damp, moisture is seeping through walls. Ventilation or a dehumidifier might solve the problem, as will correct insulation and a vapor barrier. (For patching, see Concrete.)

FRAMES: Picture

To Restore a Picture Frame:

- Use BB shot if you've got beading missing on a gold-leaf frame. Apply plaster and put BBs in when still malleable. Paint same color.
- Always wash with soap and water to get all grease and dust out before you start working.
- If you are replacing plaster of paris, mix only what will be used at one time. You can't store the stuff. Use a cut-off plastic jug for mixing.
- Secondhand stores always seem to have lots of old frames. Always check there first to see if there's one that matches up with what you need.

Regilding Plaster:

- Only use bronzing powder combined with banana oil or bronzing liquid. Use several tablespoons of bronzing powder and add banana oil a little at a time.
- Protect a gilt frame by rubbing baby oil over it and polishing with a soft cloth.

Unfinished Frames:

- Stain with liquid shoe polish in the right color. Coat, allow to dry, add another coat. Then wax.

To Remove Old Finish:

- Soak frame in a laundry tub filled with water. Leave for several days. To speed the process add a little warm water (never hot). Once the outer layer is ready to peel off, remove.
- Brown shoe polish makes an excellent finish.
- Always end up with a layer of clear shellac and wax.

To Clean Frame:

- Cut an onion in half and rub over the frame. Wipe with a damp cloth and buff with a soft dry cloth.
- Boil an onion and rub the frame with the onion water. Dry with a soft cloth. (*See also* Hanging Pictures.)

FREEZERS

- Always defrost before the frost gets to ½ in (1 cm) thick, and do it when there's the least amount of food in the freezer.
- Clean freezer once a year with a solution of 1 tbsp (15 mL) baking soda to 1 quart (1 L) of water. Dry with a hair dryer.
- Keep freezer set at 0°F (-18°C).
- Always fill from top to bottom if you have the shelf-style machine, and the fuller the better.
- Food will stay frozen for about forty-eight hours if there's a power failure, but don't open the door to check things unless it's to add dry ice or newspaper on the top.
- A properly cared-for freezer will last about fifteen years.
- Use your freezer to store stamps that have stuck together. When you need them, they'll come apart and be usable.

FREEZING FOOD

Stocking the Freezer:

- Never add more than 3 lbs (1.4 kg) for each cubic foot (30 cm^3) of space—it'll raise the temperature too much.
- Keep new packages against the side, except for baked goods, which attract moisture.
- Leave enough space for expansion in the top of containers when you store food in the freezer: ½ in (13 mm) for 1 cup (250 mL); 1 in (2.5 cm) for 1 pint (500 mL) and 1½ in (3.75 cm) for 1 quart (1 L). The wider the opening, the less space you need to leave.

Refreezing Food:

- You can refreeze if the food is cold and still has ice crystals on it, or if it's completely thawed and still smells fresh.
- Don't refreeze ice cream, seasoned food or prepared dishes.

Food That Can't Be Frozen:

- Don't freeze hard-cooked eggs, mayonnaise or juicy stuff like pears, melons or salad greens.

Food That Can Be Frozen:

- Freeze soup stock by pouring into ice cube trays and then transferring to freezer bags. It will last for three months.
- Freeze small amounts of leftover mashed potatoes in ice cube trays. Once firm, store them in your freezer and use to thicken soups and stews, with no thawing necessary.
- If a recipe calls for egg yolks, freeze the white in ice cube trays. They'll last for a year. One cup (250 mL) equals seven or eight eggs.
- I freeze enough tomatoes for the winter the following way: Cut out the green tops from freshly-picked tomatoes, dip them into boiling water for a few seconds. Dry on cookie sheets so they don't touch each other. Then put in freezer. Transfer to freezer bags when they're solid. Take out as needed, and skin them by holding under the hot water tap for a few seconds. The skin's too tough and tasteless to leave on at this stage.
- I also put any leftover tomato in the freezer for use in stews.
- Pesto sauce freezes beautifully. Just leave a layer of oil over the top.
- Parsley, dill, sage—in fact, most herbs—freeze successfully. Wash, put in a plastic container and freeze. Snip off just what you need as you go along.
- After squeezing the juice out of lemon, freeze the rind for future use.
- Freeze bananas that are about to go brown, or peel and beat, put in a container and freeze. Use them in baking.
- Cranberries, raspberries, blackberries: pat dry and freeze on cookie sheets, then bag until ready to use.
- Marshmallows keep well in the freezer.
- Always keep popcorn in the freezer.
- Clams and oysters: wash, then place in freezer for an hour and they'll open up readily.
- Make sauces ahead of time. White sauce: 1 cup (250 mL) butter and 1 cup (250 mL) flour mixed together, pushed into ice cube tray and then cut into cubes. When you need to make the sauce, add one cube to 1 cup (250 mL) milk.
- Do a tomato-based sauce when vegetables are fresh: chop onions, green peppers and tomatoes. Cook to reduce moisture. Add garlic, herbs and spices during this process. Then bag and freeze. Finish cooking when needed.

Cooking Frozen Food

- Always cook packaged vegetables to your own taste (al dente or crunchy is best), not what's on the package—they always overestimate cooking times.
- To cook frozen vegetables, pour boiling water over them and then rinse. Cook as you would fresh vegetables.
- Defrost meat and poultry in the fridge and make sure it's completely thawed.
- Heat sauces and casseroles in a double boiler.
- When you cook frozen food, always add a little something fresh, whether it's lemon juice, butter or parsley to give it some zip.

FROGS AND TOADS

If you are lucky, a frog or toad may take up residence in your garden. They eat all insects, including revolting slugs that may attack your plants. They will need water and a place to live, such as an overturned clay pot.

FROZEN LOCKS

- If you can't get the key into the door of your car, light a match and hold it to the end of the key and it will probably slip in easily.
- Plug a hair dryer into a plug in the garage and aim at the lock. It'll only take a few seconds.
- Carry a drinking straw with you. Put one end in the lock and blow warm air into the lock.

FROZEN PIPES

If pipes freeze during a cold snap, do the following:

- Plastic pipes: use a hair dryer to blow hot air on them.
- Metal pipes: open nearest faucet, then use a heat lamp, hair dryer or propane torch to warm the pipe.

FRUIT (See Bananas; Freezing Food; Pesticides: On Produce; Ripening Fruit and Tomatoes.)

FURNACES

Keep your furnace running smoothly to save energy.

- Once a month, hold the furnace filter up to a light bulb. If you can't see light easily, clean it by vacuuming, or replace it.

- You can save money if you turn off the pilot light of a gas furnace during the summer.
- Vacuum the area around the furnace annually. If you have a gas furnace, be sure there is no gas odor—sucking any gas into the motor of your vacuum cleaner could cause an explosion. Every two years you should have the furnace checked by a professional; if you have a high-efficiency furnace, have it checked annually. Call your local utility company for recommended service companies. If you have an oil furnace, have this done every year.
- Set the thermostat at 68°F (20°C) and, if you feel a little chilly, put on a sweater. You certainly won't want it warmer than that at night. Try getting along with 61°F (16°C). If you can afford it, put in a programmable thermostat that automatically turns the heat down at night. By setting your furnace any higher, your heating costs rise about 5 percent a degree.
- A furnace loses up to 40 percent of its heat through the chimney (a high-efficiency furnace loses 8 to 10 percent). If you have an oil furnace, consider installing a heat reclaimer. It's a small fan that blows air through the reclaimer back into the house. Don't use one on a gas furnace.
- Make sure the oil burner on an oil furnace is operating efficiently by having a heating contractor inspect it.
- If it's time to install a new furnace, try to get one of the new energy-efficient furnaces with heat exchangers built-in.

FURNITURE CARE AND MAINTENANCE

It isn't necessary to use spray products for polishing furniture. If you do use them, be sure to look for a message that says it is an ozone-friendly product—even so, you are contributing bulky waste to the garbage.

Keep the following in mind when you are polishing furniture:

- It is always more effective to polish furniture on a warm dry day. Moisture in the air will impede a high gloss.
- Apply polish that's been warmed by standing it in some hot water. Do not put store-bought spray cans of polish in hot water—they could explode.
- Always use polish that matches the finish rather than the wood. For high luster use liquid polish, oil finish or paste wax. Rubbing is the secret of a good shine.

- Remove wax buildup by mixing together one part white vinegar with one part water. Dampen a soft cloth with the solution and gently rub into the surface. Dry immediately with a soft, dry cloth.

Furniture Polishes:

- A good all-purpose furniture polish for quick polishing jobs is olive oil. Place a little oil on a damp cloth and rub into the surface till dry.
- For unprotected furniture, as well as furniture that's been varnished, lacquered or shellacked, mix one part lemon juice with two parts olive or vegetable oil. Store this in your own spray bottle. Heat the bottle in a pan of hot water, then apply and rub dry with a soft cloth.
- For antique furniture that's fading, try clear or brown shoe polish.
- For teak furniture, dust regularly with a few drops of lemon oil on a soft cloth; wipe occasionally with a damp sponge and a little detergent on the sponge to remove finger marks and grease spots.
- If you dump furniture polish on a carpet, soak up with rags that you can reuse. Sponge clean with liquid detergent and rinse with clear water.
- To remove furniture polish from clothes, sponge with a liquid phosphate-free detergent and rinse in clean water. Sponge out any remaining stains with methylated spirits.

Removing Stains on Furniture:

- Since any alcohol spill could wreck a finish, wipe up immediately.
- Glass rings can be removed by a mix of mayonnaise and white toothpaste. Wipe dry and polish.

(*See also* Leather; Stain Removal: Fabrics.)

FURNITURE: Finishing Wooden Furniture

Unpromising and unpainted furniture can be transformed the following ways. These methods absolutely do not apply to any quality furniture—use them only on rustic pieces.

- A shellac finish: combine 1 cup (250 mL) of cut shellac with 3 cups (750 mL) of shellac solvent and mix well. Apply with a clean dry brush. Dry and sand lightly. Repeat for seven layers, then wax with Butcher's paste wax, Johnson's paste wax or Mm-Wax.

- An oil finish: combine ¼ cup (60 mL) boiled linseed oil, ¼ cup (60 mL) raw linseed oil, ½ cup (125 mL) turpentine and varnish of your choice, and 1 tsp (5 mL) shellac. Rub in with clean cloth. Let sit ten minutes. Do this from three to six times, the more the better.

FURNITURE REFINISHING

- Rub in furniture refinisher with #0000 steel wool one spot at a time, overlapping until the surface is done. To remove the streaks, go over again with clean steel wool along the grain of the wood, then wipe with a soft cloth. Dry, then buff very gently with #0000 steel wool.
- If you use a stain, brush or wipe it on gently, let sit for a few minutes and wipe off. Repeat until you get the right color.

FURNITURE REPAIRS

- For stains you want to refinish, first test by working a small area on the bottom of the object.

Burns:

For solid wood pieces, scrape the burned area with the tip of a knife or a piece of fine sandpaper and fill the hole with polish or wood filler and resand. Or you can use a mixture of glue and sawdust to fill the hole.

- Rub mayonnaise into the burn and let sit before wiping off.

Cigarette Burns:

- For superficial burns, combine rottenstone (a crumbly limestone usually used to polish metals) with sewing machine oil and rub gently. Wash well with mineral spirits afterwards.
- For deep burns, scrape away the charred wood with a razor blade or an X-acto knife before filling the hole with wax stick to match the lightest grain of the wood—it will darken with time and use. To apply the wax, heat a grapefruit knife, then hold it against the stick of wax and work the melting wax into the hole. Overfill the hole so you can scrape it level once the wax is cooled. This treatment is appropriate for a small area only.

- To simulate a wood grain, dip the teeth of a comb in a darker stain chosen to match the existing wood. Run it lightly over the patch to create a grain.

Decals or Paper Stuck to Furniture:
You would be well advised to approach a professional if decals or stuck paper are your problem. It can be very tricky to remove them without damaging the furniture.

Dents and Gouges:
- Put a damp rag over the spot and press the tip of a warm spatula to the area. The swelling should disguise the dent. Or use a steamer (a travel steamer will serve very well) directly on the dent. Smooth surface by using a fine sandpaper.
- Melt a wax stick or a crayon by tying it to a pencil and holding it over a candle flame. Apply to dent, cool and shave off excess.

Glue:
- Rub spilled or excess glue with peanut butter or salad oil. In a pinch try some old fashioned cold cream on it. (*See also* Glue and Gluing.)

Holes or Cracks:
- Use children's crayons to get a specific color in a tiny hole. Press the wax into the surface and scrape it flush with the edge of a knife. Sand and seal with a coat of French polish before finishing with a matching varnish.
- Use a shellac stick by melting with a smoldering iron. It sets immediately and can be evened out with a sharp blade, then sanded.

Scratches:
- General rule of thumb: try hiding any scratch by applying paste wax of a slightly darker color with very fine steel wool, then buff.
- For fine scratches, rub a mixture of equal amounts of lemon juice and vegetable oil into the scratch and polish.
- Light woodwork: fill with beeswax to which some color has been added.
- White woodwork: fill with white shoe polish.
- Deep scratches in dark woodwork: drip colored sealing wax deep enough to fill, smooth and then polish.

- Superficial dark scratches: touch up with iodine or shoe polish, then rub with #0000 steel wool.
- Shoe polish works wonders on any number of places: use a soft cloth to apply the wax or paste type. If it's light, use brown; if it's black lacquered wood, use black polish; mahogany needs cordovan. Then buff to polish.
- Iodine is useful on red finishes.
- Use children's crayons in the right color or get a professional wax stick to match the wood finish.
- Walnut: use the meat from a walnut and rub it into the scratch. Wipe with a dry cloth.
- Mahogany: use a child's crayon in the same color or brown paste wax.
- Blond finishes can be touched up with tan shoe polish.

White Rings:

- White rings are usually made by heat or water marks.
- Rub gently with toothpaste and buff away the white.
- For more stubborn stains, add baking soda to the toothpaste. Buff with a soft clean cloth.
- If the surface has been made rough by water damage, apply cleaner with #0000 steel wool, rubbing in the direction of the grain and repolish with paste wax.

FUSE BOXES

- Always replace a fuse with one of the same amperage. Don't cheat on this—it isn't wise.
- Keep spare fuses handy to the fuse box or inside the fuse box, if there's room.
- Wear rubbed-soled shoes and never stand in a wet spot when handling fuses in the fuse box.
- Keep the fuse box locked if you have small children in the house. In some regions, it is a legal requirement that the fuse box be located about 4 ft (1.2 m) from the floor.

GARBAGE

Garbage is one of the major impending disasters in North America. We produce more than anyone else in the world and we're running out of room to stack it. It's imperative that everyone reduce waste. It's recommended that a family of four should only produce about the equivalent of one garbage can a week.

- Instead of using garbage bags, use cans instead.
- Spray garbage cans with household ammonia or sprinkle hot cayenne pepper on top to keep animals away.
- Hose down garbage cars regularly, but when you need to do a thorough job, give a shower job in the bathroom. Use hot water, detergent and a strong brush. Tip upside down to dry.
- If you are stuck with a garbage disposal (they waste water and electricity), grind up some lemon or orange rinds and add to the garbage disposal if it smells.

(*See also* Bags; Compost; Recycling.)

GARDENS

- Use half orange or half grapefruit skins to plant seeds in. Keep moist and when ready, plant seedlings, skins and all. Citrus adds nutrients to the soil.
- To keep cats and dogs from urinating, among other things, on your plants, put a few garlic buds and three to four red hot peppers in the blender. Drop this mixture, along with a few drops of biodegradable dishwashing liquid, in a bucket of water. Sprinkle around the edge of the garden.

- Black or hot cayenne pepper sprinkled around shrubs will deter the raised leg.
- Grind up citrus and put it near any plant the cat seems attached to—they don't like the smell or taste.
- To keep the blades of the lawn mower from getting grass stuck to them, wipe with vegetable oil before you mow.
- Keep garden tools free of rust by pushing them into a bucket of sand mixed with ¼ cup (60 mL) oil (vegetable oil will do). It will scrape off the soil and coat the tools. Also keeps edges sharp.
- Store the hose inside an old tire and hang it on the garage wall. You can get about 100 ft (30 m) of hose in there.
- If you have a new hose, put fake patches and tape around it in a few spots—it would be much less likely to be stolen.
- If you have an old hose, bung holes all over it and you've got yourself a new dribble hose.

(*See also* Companion Planting; Insect Control; Lawn Mower Maintenance; Lawns; Weeds.)

GARLIC

- For a light garlic flavor, don't pierce the clove after peeling, sauté whole clove in oil and remove before cooking.
- Hard-to-peel garlic will shed its skin more easily if it's dropped in boiling water for a few seconds.
- The finer garlic is chopped and the darker it is allowed to become in the cooking, the stronger the flavor.
- Make garlic oil for dressing by leaving cloves in the oil for a day, or put cloves in vinegar and leave for a week. Then discard the garlic.
- Store garlic in a small open container. It won't transfer its odor to other foods.
- The best way to crush garlic is with a mortar and pestle.
- A magnificent use of garlic is to cook a chicken with forty peeled cloves and serve them along with the fowl. They will taste sweet and delicious.

GATE REPAIR

- A sagging gate may mean that the brace is broken. Remove the brace and take out any rusty bolts. Shear off rusty bolts with a saw and replace with a new brace.
- To test for rot: push a penknife blade into the wood. If it goes in easily you'll have to replace it.

- If the gate is old and the posts haven't been preserved below ground level, consider replacing them. Set them in metal post holders that you drive into the ground. Or set them in concrete to avoid the possibility of the ground heaving in the spring and fall—the metal post holders will shift.
- Always replace bent hinges.
- To slow corrosion, roll each screw up and down a candle. Tip heads with wood filler.
- If screws insist on pulling out of the wood, remove and fill holes with a glue and sawdust mix. Reset the hinge when the glue is thoroughly dried.

GERBILS

Gerbils are intensely active, then out cold. Most of their activity takes place during the day.

- To pick up a gerbil, put something tempting on your palm and it will come out and climb on your hand. Don't let it fall. Use your other hand as a barrier. If it escapes, do the same thing. Never pick it up by the tail.
- Make a cage with solid sides up to 8 in (20 cm) so that litter doesn't scatter. Use 3 to 4 in (8 to 10 cm) of litter—wood chips, sand and soil, burlap or sawdust. Gerbils like to make tunnels, so keep this in mind and make sure it's deep enough. A clean empty can will create that illusion and provide a nest.
- One gerbil won't make enough mess to change the litter more than once a month.
- A brick will provide an eating platform.
- Feed ⅓ oz (10 g) to ½ oz (15 g) of rodent pellets a day. Add nuts, fruit or vegetable as a daily treat. It might eat its own feces. This is normal.

GLASS: Cracked

- A hairline crack can be fixed with sodium silicate. Dilute the liquid with an equal quantity of water and paint it on the crack. Wipe and then let dry. Reapply.
- Call china shops and find one that will repair chipped glass.

GLOVES

Save a bit of money and clean your own leather gloves:

- Wash in cool water with a capful of Forever New (my favorite cold-water laundry soap). Rinse, put the gloves on your hands and squeeze as much water out as possible. Dry on a towel.

- When leather gloves are almost dry, put them on and stroke them into shape. They will look much better that way and will keep them pliable.
- To keep them supple, rub with a bit of the leather conditioner you use on boots.
- To mend gloves, put a marble in the finger as a darning base. Most needle cards have a needle large and strong enough to deal with leather.

(*See also* Leather.)

GLUE AND GLUING

Keep the following principles in mind:

- Never apply new glue to old. Scrape old glue off and sand the surface.
- Hot vinegar will sometimes soften old glue.
- Apply glue to both surfaces to be joined.
- If glue drops on any surface, rub the following on: salad oil, cold cream or peanut butter.
- To clamp two pieces together, wrap several turns of rope around the join.
- The safest glues to use are carpenter's glue and water-based white glue. Some others contain toxins and are very damaging to the atmosphere.
- When you are working with glues that aren't water-based, be sure that your whole house is well ventilated. Always read the labels and instructions carefully.

GOLD

- If it tarnishes, it isn't pure gold. A very weak ammonia solution will clean it.
- Polish real gold with a solution of pure soap and water to remove grease or dirt.

GRAVY

- The best gravy is made with the most commercial white flour. Sprinkle it directly on the drippings and cooked bits in the bottom of the pan. Add water or the potato water and use a wire whisk to avoid lumps.

- If you add wine to gravy, cook for at least fifteen minutes to boil off the alcohol. All you want is the flavor.
- For dark, rich-looking gravy, brown the flour gently in a pan first. Add flour in a separate Pyrex dish when you put the roast in the oven. It will be brown by the time the roast is cooked.
- To remove excess fat, let the fat rise to the top and pass a piece of extremely dried-out bread over the surface to absorb it.
- A bit of baking soda will cut the grease in gravy.

GROUT: REPAIR

- Clean out an old plastic glue bottle and fill with grout. Now you'll be able to squeeze it out in smaller dabs and the grout won't dry out as quickly.

GUTTERS

- Check gutters every year to prevent seepage into the house.
- Check the brackets holding gutter to the roof to make sure none are loose. Tighten the brackets to keep perfect pitch.
- Keep gutters free of leaves and rubbish. It's easy to remove most of this by hand and then flush out with a garden hose. Do this after all leaves have fallen. Check again in spring. The detritus can go into the compost.
- Repair rust with a piece of heavy canvas or fiberglass patching. Seal with roofing compound, which is available in tubes.
- To protect leaves from clogging downspouts, put a piece of wire over the opening.

HAIR CARE

- Check labels to see if the product is biodegradable.
- Avoid hairsprays sold in an aerosol can. Use gels, setting lotions or pump-sprays instead.
- To make your own gentle-hold hairspray, try the following. Boil one chopped lemon in 2 cups (500 mL) water for ten minutes. Strain and store in a pump-spray bottle in the refrigerator.
- Buy products with the least amount of packaging.
- Write to the manufacturers of overly packaged or non-recyclable packaged items.

HANDBAGS

- Mesh bags: rub with a soft cloth dampened with vinegar. Sprinkle ammonia on another cloth and rub. Wipe clean with clear water. Dry and buff.
- Straw bags (natural): beat the white of an egg and add juice of one lemon and 1 tsp (5 mL) Epsom salts. Brush the mixture on with an old toothbrush. Wipe with damp cloth and dry in the sun.
- Straw bags (colored): do the same but exchange water for lemon juice. Or use a cloth with a bit of toothpaste on it to spruce up the handbag.
- Clean leather with saddle soap.
- Remove marks with clean pencil eraser.
- An art-gum eraser will get marks off suede.
- Scratched leather bag can be renewed with a soft cream-type shoe polish. Buff thoroughly.
- If you must patch a leather bag, use a kit from the craft or hobby shop to do the job. If the patch is obvious, shape it so that it becomes an ornament or decoration.

- If a clasp won't close, bend the frame slightly to reset the alignment.

HAND-WASHING CLOTHES

The safest way you can clean your clothes is by doing your own hand-washing. You'll use less water and no chemicals, plus it's cheap.

- General tip: mix soap and water thoroughly before adding clothes.
- Drip-drying: after washing, hang on rustproof or padded hanger and blot with a towel, but don't squeeze dry. Straighten garment and smooth with hands. My all-time favorite soap for anything I care about is Forever New. It's just soap, but it works like magic.
- Silk: dry cleaning can discolor delicate silk blouses. Soak silk garments in two capfuls of Forever New (or other cold-water soap for delicate fabrics) and cold water to cover for about five minutes. Swish gently and check for stains. If they haven't disappeared, repeat. Then rinse and drip-dry.
- Or try washing silk in warm water with a pure soap such as Ivory. Add 1 tsp (5 mL) methylated spirits to revive the sheen.
- For difficult stains, pour Forever New directly on spot and rub in very, very gently. Then proceed as usual.
- Knits, wools: use cold water and gently squeeze soap through the garment. Rubbing transfers stains so keep layers away from each other. Rinse but don't wring dry. To dry, roll in towel and blot. Lay flat. Block to original shape.

HANGING PICTURES

- Hang fishnet along one wall of a child's room and pin artwork to it instead of the wall.
- Hang pictures at eye level where they are intended to be viewed most often.
- Use brown paper cutouts to experiment with before hanging pictures, and tape to wall to get right distances. Once it's right, stick pencil through paper to mark the spot.
- A good method of hanging a wall of drawings or photographs is to lay them out on paper on the floor and move them around until you get a pleasing pattern. Then mark where you want to hang them on the paper and hang the paper on the wall. Put in the picture hangers in the right spot. Remove the paper and hang the pictures.

- If you haven't any picture hangers in the house, sewing machine needles driven into the wall will hold up to thirty pounds.
- On wallpapered walls, cut a V in the paper where you want to put the nail. Peel back the paper. If you change your mind, glue the paper back in place.
- Mark spot on wall with a wet finger mark. It will dry without leaving a trace.
- Before hammering in nail, cover spot with tape to keep plaster from crumbling.
- Put masking tape on the corners of a picture to keep them from marking the wall.
- For heavy pictures, loop wire back through the screw eye and wind it around the main wire for more strength.
- Hang a large picture by hooking one end of the wire to the hook and sliding it along to the center. Adhesive tape in the center of the wire will keep it from slipping.
- To attach an unframed picture, dab toothpaste on each corner and let dry on the wall. When you take it down, just wipe off the toothpaste.
- If you have nail holes you want to disguise, use toothpaste to fill them in.

(*See also* Frames: Picture.)

HARD-WATER DEPOSITS

- Get the scale out of your kettle by adding ½ cup (125 mL) vinegar and fill with water and boil. Repeat three or four times every three months.
- For hard-water stains, rub with a pumice stone. Or cover with rags soaked in vinegar. Let stand for an hour, remove and rinse.
- For darker stains that won't come out with other methods, do the above, only use bleach instead. Then scrub with cream of tartar on a damp cloth.

HEAT CONSERVATION

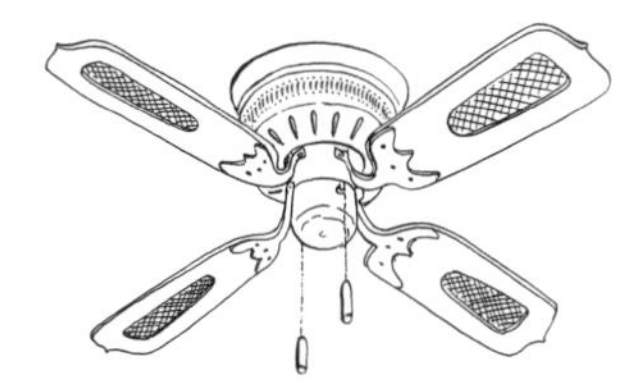

- Install a ceiling fan to force warm air down from a cathedral ceiling and stairwells in winter.
- Plug the opening of unused fireplaces and chimneys.
- Use glass doors on fireplace openings to reduce heat loss.
- Make sure the damper is closed when fireplace is not in use.
- The flue on a fireplace should be one-tenth of the whole opening for greatest efficiency.

- The correct placing of foundation plants can cut your heating costs considerably. They act as insulation when placed about 2 ft (0.6 m) from the foundation of the house. Make sure they buffer your house from prevailing winds.
- Put heavy aluminum behind water radiators to deflect heat into the room instead of the walls.
- Put the thermostat and baseboard heaters away from both drafts and sources of heat.
- Close registers in rooms that aren't being used in winter.
- Don't leave bathroom exhaust fans on for more than a few minutes.
- If hot air is escaping from basement heating ducts, use duct tape to stop it.
- Clean or replace furnace filters every two months.
- Use a small electric heater to warm up bathrooms rather than turning up the thermostat and heating the whole house. But don't use them unless they are very safe and you don't have small children around. Turn the heater off as soon as the room is warm.
- A heat pump exchanges inside for outside air. In warm weather it will take warm heat from the interior of the house to the outside, and bring warm air from the outside in cooler weather. A good heat pump can save as much as 65 percent of your heating needs.

HEATERS, ELECTRIC (See Electric Heaters.)

HERBS

- Test herbs for potency by rubbing between fingers. One tsp (5 mL) of fresh herbs equals ½ tsp (2.5 mL) dried.
- Fresh herbs can be frozen. Wash and put in an airtight container. Use straight from the freezer; just snip off as much as you need each time.
- Don't keep dried herbs for more than four months.
- If you double the recipe, don't double the herbs—just a little more will suffice.
- To make a bouquet garni, put thyme, marjoram, bay leaf and parsley in a metal tea holder.

HOME OFFICES

An increasing number of people work at home, whether they are telecommuting or have their own home-based business. Those who have this advantage are already helping the environment in one way: not having to sit in rush hour traffic, thereby preventing car pollution. Other "green" options for your home office:

- Try to take advantage of natural light, which is healthier for you and will cut down on the electric bill.
- If you have the option of a few rooms in your house, try to pick the one with the best natural light. Also consider installing windows or skylights.
- If you do need artificial light, use energy-efficient fluorescent bulbs.
- Turn off lights and equipment when you leave the room.
- Buy a plain-paper fax machine rather than one that uses chemically-treated thermal paper. An extra bonus is that thermal paper fades and yellows over a very short period of time, unlike plain paper.
- Recycled paper for your printer, fax machine or photocopier isn't the only environmentally friendly office supply you can buy. Look for water-based magic markers, non-toxic correction fluid, stapleless staplers and recycled clipboards, too.
- Keep a scrap box for paper that has only been used on one side—it makes great scratch paper.
- Use "green" packing materials, like recycled padded envelopes, cornstarch, peanuts and shredded newspaper. Some shipping chains will take back styrofoam packing material, so call around for details.
- Office equipment can contribute to indoor air pollution. Placing a few houseplants in your home office will help neutralize pollution and boost the room's decor. If you're not much of a plant person, you may want to consider getting an air filter, or opening a window if you live in a low-pollution area.

(*See also* Computers.)

HOME REPAIR

- Always turn off the electricity or water, or both, before you start to work on anything around the house.
- Check for the most obvious problem. It's embarrassing to call in the serviceman for a washing machine that won't fill, only to find that the tap at the head has been turned off.
- Lay out all the tools you need before starting a project.
- Read the instructions on the package first to find out what you'll need. As well, read the instruction manual for the item you're repairing.
- Lay out parts in exactly the same order as you remove them. Keep a diagram if you can and mark which parts go where.

- An easier method is to use a magnetic strip (the kind that holds knives) to put all the pieces you are working with in order. This way they won't get misplaced or lost.
- Always check to see if the warranty is in effect—it's a good idea to keep them all in a file or a drawer used especially for that purpose.
- Get organized: keep home files with instructions for appliances; make a list of repair services, workers and receipts.

HOSES (See Gardens.)

HOUSEBREAKING PETS

Cats:

- There is biodegradable kitty litter on the market now. Look for it.
- Have a kitty litter pan and keep it clean by pouring a box of baking soda on the bottom before you add kitty litter. It will keep it smelling nice.
- Clean a used kitty litter pan with vinegar.
- Just introduce the kitten to the pan, leave the door to the room closed and it will get the idea in a day. A sharp "No!" if it tries to go outside the pan and a bit of praise for doing the right thing are enough.
- Plastic dishpans make the best litter boxes.
- With well-deodorized litter, you need only lift out the feces, take a sniff of the pan, and if it smells okay to you, it will smell okay to the cat. It's not necessary to change the litter until it starts to smell.

Dogs:

- Don't rub the dog's nose in its own feces if it has an accident. This may encourage it to eat the stuff as it grows older. Scold only when you catch it in the act because dogs don't have much memory.
- Take the animal outside as soon as it's been fed and encourage it to go outside.
- The animal will signal that it has to go by whimpers, whines, squatting and then action. Catch it somewhere in the act and get it outside immediately.
- Or paper train by setting the animal on the paper after each meal. Lavish praise when it goes. Show it that you aren't pleased if goes off the paper.

- Clean up immediately. The cleaner you are, the cleaner your dog will become.
- Make a den out of an old crate with paper, water and a favorite blanket. Keep the puppy in there when you aren't around and it will learn to hold its bladder. It will keep it clean.

HOUSEHOLD HAZARDOUS PRODUCTS

Household hazardous products (HHPs) are a danger to us, the water we drink and the air we breathe. Some examples you may find around your house include antifreeze, batteries, chemical strippers, contact cement, drain cleaners, flea collars and sprays, insecticides, paints, spot removers and oven cleaners.

In addition to the harmful effects they may have on us, HHPs are harmful to the environment, particularly when improperly disposed of. If HHPs are poured down the drain, they end up in our rivers, lakes or oceans because water treatment plants can't treat them. If they're thrown away in the regular garbage pickup, toxic materials end up in our soil and groundwater.

To lessen the dangers, we need to reduce our use of HHPs, store them properly and find alternatives.

Reduction:

Ask yourself the following questions when you go shopping:

- Do I really need to buy this? After all, a "new, improved formula" may not be any better than an old tried-and-true method.
- How much do I need? Only buy what you need to finish the job.
- How do I get rid of this when I'm finished? Find out where your local hazardous waste disposal facility is. Some communities have special collection days—watch for them.

Safe Storage:

- Store HHPs in safe places out of the reach of young children and pets.
- Ensure the storage place is well ventilated and fire-safe. Storing products near a heat source could cause them to explode or ignite.
- Make sure the containers aren't broken and keep them securely capped or sealed.
- Keep HHPs away from appliances with open flames, e.g., gas furnaces and clothes dryers.

Finding Alternatives:
(See Batteries; Clothes Washing; Commercial Cleaning Products; Insect Control.)

HOUSEPLANTS

- Water plants in the morning. Use room temperature water that you've left out overnight so that any chemicals have evaporated.
- For plants with heavy water demands, let the surface dry out. For those with moderate water demands, let 1 in (2.5 cm) of soil dry out. And for those with light water demands, let two-thirds of the soil dry out.
- Indoor plants thrive in temperatures of 65°F (18°C) to 82°F (27°C) degrees. They don't like drafts or cold windows.
- Indoor plants will provide you with another source of humidity if you place them together on a bed of pebbles that is kept moist.
- Never put gro-lights closer than 3 to 12 in (8 to 30 cm) above the plants.
- Clean plants with a feather duster, not milk. Milk clogs the stomatae which helps them breathe.
- African violets sprout more flowers and look better by putting in a couple of rusty nails—they need iron too.

When You're Out of Town:

- To take care of your houseplants when you're out of town, fill the bathtub with a few inches of water. Stand bricks on their sides and put the plants on the top.
- Or put them on thick towels and a few inches of water and they'll absorb what they need.
- Run a clothesline rope or heavy twine from a raised pail of water into the soil of the plant.

Plant Tonics:

- Every few months try a shot of Geritol or 1 tbsp (15 mL) of castor oil followed by watering.
- Or dissolve 1 tbsp (15 mL) gelatin in 1 cup (250 mL) of hot water. Slowly add 3 cups (750 mL) cold water. Water plants with this during the growing season.
- Another lift to plants will come from that stale club soda sitting around in the fridge—once its been warmed to room temperature.
- Use very weak tea to water a sad-looking fern, then add the wet tea bag to the soil.

- Save the water from boiling eggs and top up plants with this. Or put eggshells in water for a day and strain.
- In the dead of winter, put your houseplants under a fluorescent light for several hours at a time.

Pests:

- A wooden kitchen match, inserted into the soil sulphur side down will get rid of worms in ferns. (*See also* Insect Control.)

HUMIDIFIERS

A humidifier that is clean will use energy more efficiently.

Humidifiers Attached to the Furnace:

- Recycle pads by soaking them in vinegar for an hour. Do this at the beginning of each season.

Free-standing Humidifiers:

- Wash the drum pad in warm soapy water once a month. Scrub out the mineral buildup in the water container at the same time.

Ultrasonic Models:

- Wipe nebulizer with a product made for the humidifier, using a soft brush, then rinse. Every time you have to fill up the machine with water, rinse it out with room-temperature water before you fill it up again with cold water. Use great care, as the nebulizer breaks easily.

HUMIDITY CONTROL

Keep house well ventilated. Close windows on humid days; open when the air is dry.

- If the closet or room is damp, install a vent, louvered door or exhaust fan—or all three.
- Leave thermostat at 65°F (18°C) most of the time.
- If you must place furniture near a vent, place a pan of water under it to keep the wood from drying out.
- Put a pan of water under radiators. Add a piece of charcoal to keep the water sweet.
- To protect a grand piano, place small containers of water on the sounding board.

- Plants will do more to keep your home humidified than anything else. Just make sure you keep them watered.
- To add a bit of humidity to a dry winter house, instead of emptying the bathtub immediately, let it stand for an hour or so until it's cold. It will also heat up the air.

ICE REMOVAL

Salt is corrosive and damaging to the environment as well as boots and carpets. Try these alternatives:

- Shovel snow immediately after a snowfall.
- To provide some traction, sprinkle sand or gravel over icy areas.
- You can also use biodegradable kitty litter.
- Granular lawn fertilizer will also melt ice, and it won't burn the grass or ruin rugs when it's tracked indoors.
- Calcium chloride pellets will melt ice at −25°F (−31°C). You won't need as much as you will with salt and calcium chloride is less damaging.

INSECT CONTROL

Don't get out the killer chemicals because you've got a few pests in your house and garden. Use environmentally-friendly alternatives. About 99 percent of what we think of as pests aren't harmful.

In Your Home:

- The best way to prevent bugs is to deny them of their basic needs: food and water. So be sure to store food in airtight containers. Place garbage in a container with a tight lid and take it out every day. Keep counters, floors, cabinets, shelves and stovetops clean. Fix leaky pipes, as this is where insects often get their water supply.

In Your Garden:

While many people see all insects as pests, there are some that are very useful for keep down the ravening hordes of pests attacking your garden and home.

Here are the major ones:

- Lacewings are the most beneficial bugs of all for controlling aphids. They are attracted by members of the carrot family, wild lettuce and oleander.
- Ladybugs attack aphids, leafworms and leafhoppers. Plant angelica, evergreen euonymous, goldenrod, morning glory and yarrow to attract them.
- When they are young, praying mantis feed on aphids and leafhoppers; they eat chinch bugs, crickets, beetles and caterpillars when they are older.
- Trichogramma wasps attack two hundred pest species, including the dreaded spruce budworm. They do not sting people. They are attracted by members of the daisy and carrot family, especially Queen Anne's lace, goldenrod, oleander and strawberries.

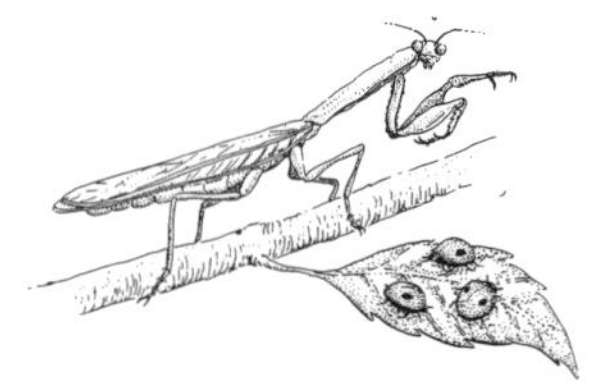

In addition:

- Animals such as toads will keep a garden very clean.
- Set up a bird-feeding station to attract birds that eat insects you don't want.

Do-it-yourself Repellants and Insecticides:

- Cut up large rhubarb leaves and boil for three-quarters of an hour; cool and strain. Then spray plants. If you spray vegetables with this mixture, be sure to wash them first before eating; rhubarb leaves are poisonous.
- Also, ¼ cup (56 g) dried basil added to 3 quarts (3 L) water is nontoxic plant spray.
- Collect ½ cup (125 mL) of the offending bugs and grind them up in an old blender reserved for this purpose, then add them to 2 cups (500 mL) of warm water. Strain and make a liquid with ¼ cup (60 mL) juice to 2 cups (500 mL) water. Reapply to both sides of plant leaves after it rains. Either the smell of death gets them or it may attract their natural enemies.
- Diatomaceous earth is a fine powder made of ground-up fossils of ancient creatures. Use a ring of it around a plant or make a spray with 1 tsp (5 mL) insecticidal soap in 5 gallons (20 L) water to which ¼ lb (113 g) of DE has been added.

- Chop up anything that smells strong such as garlic, onions, chives and add to water. Strain and spray.
- Crush three cloves of garlic and blend with 1 tsp (5 mL) hot cayenne pepper and 1 quart (1 L) water. Let stand for ten minutes and strain through nylon sock. Dilute one part of mixture with four parts water and spray plants.
- Or combine 1 gallon (4 L) water with 2 tbsp (30 mL) fresh garlic juice (not powder), 1.1 oz (32 g) diatomaceous earth and 1 tsp (5 mL) rubbing alcohol. This mixture can also be frozen.
- Combine 2 tbsp (30 mL) pure soap with 1 quart (1 L) water. Don't use detergents.

Commercial Products and Ingredients:
- Bacillus thuringiensis (BT) is one of the most beneficial of all natural powders. Mix with insecticidal soap or a seaweed solution to make a spray. Dipel, Thuricide, Envirobac and Fossil Flower are brand names containing BT.
- Fossil Flower, Insectigone, Safer's Roach and Crawling Insect Killer all contain diatomaceous earth.
- There are some good insecticidal soaps on the market—Safer's is one of the best. It breaks down in a week or two, so you must reapply and it works only on direct contact with the insect.
- Rotenone and pyrethrum are two natural insecticides that are generally available. Atox, Deritox and Wilson's Vegetable and Garden Spray contain them. They can be used on vegetables to control aphids, potato beetles, leafhoppers, cabbageworms and corn borers.
- Garden sulphur treats mildew, rust, black spot, scab, black knot and some mites.
- Thuricide is an organic insecticide that acts on leaf-chewing worms and caterpillars.
- Tanglefoot is a sticky substance applied around tree trunks to keep insects from the foliage or buds.

(*See also* Companion Planting; Household Hazardous Waste.)

INTERNET SAFETY

The Internet is a wonderful place for both adults and kids—full of information, resources and sites that are just fun to visit. However, just as with any other toy or tool, there are safety rules for Internet use. Here are ways to keep your children safe when they're online:

- Place the family computer in a public room in the house. If the only computer is in your office, monitor your kids' usage.
- Share online time by surfing with your children, and encourage them to tell you what kind of sites they found while surfing alone.
- Consider purchasing an Internet filtering program that will restrict your child's access to pages that are unsuitable for them.
- Use kid-friendly search engines, like <www.yahooligans.com>.
- Instruct your children to never give out personal information unless you give them permission. And read the website's privacy policy before allowing your child to respond.
- Be wary if your child wants to meet someone in person that they've talked to online. Make sure you are present if you allow the meeting.
- Tell your children to never respond to chats or e-mail messages that make them feel uncomfortable, and ask them to tell you when they get these kinds of messages.
- If you or your child encounter any inappropriate Internet use, inform your Internet service provider.
- If you find child pornography on the Internet, are sent child pornography by e-mail, or if you believe an adult is trying to extort information from your child, or lure them to a face-to-face meeting, inform a federal government agency, such as the FBI or RCMP. Save the e-mail message or make a note of the site's URL (which will begin with <http> or <www>). Don't download or copy any materials yourself.

INVITATIONS

Formal invitations should receive a formal reply.

- Always handwrite your reply on quality paper in ink.
- Follow the format of the invitation exactly:

> Mr. and Mrs. Andrew Smith
> accept with pleasure
> the kind invitation of
> Jonathan Buck
> to attend
> his graduation on
> Saturday the nineteenth of May

Do the same if you can't attend, inserting "Regret that they will not be able to ..."

IRONS

If you have a steam iron, don't use the water that drains from your fridge or freezer—it isn't clean enough. Use water from the dehumidifier.

Cleaning Your Iron:

- Place a brown paper bag with waxed paper on top and pass the medium-heat iron across. The iron will have smoother surface.
- Working around the holes, use toothpaste to clean the bottom of the iron, then wipe off with vinegar.
- Use fine steel wool to clean off a non-steam iron. Dip the steel wool in powdered pumice or whiting, then wipe with vinegar.
- To clean an iron while it's hot, sprinkle salt over a piece of paper, run the iron over it and continue ironing.

IRONING

- Starch an ironing board cover and back it with aluminum foil. The heat will reflect through and the cover will stay clean longer. (See Clothes Washing for a homemade starch recipe.)
- Always iron clothes lengthwise; circular strokes will stretch the material.
- Pressing means raising and lowering the iron on the fabric.
- Start with the clothes needing the lowest setting first, then move on.
- Be sure ironed clothes are dry before putting them away.
- Iron on the wrong side of the fabric to keep the fabric from getting shiny. This is especially important for any fabric with a nap.
- Never iron dirty clothes, you'll only set the stains more firmly into the fabric.

Dampening Clothes:

- Tumble clothes with wet towels and set on low heat.
- Dampened clothes can be put in the freezer until you're ready to iron them.

Ironing Tips:

- Embroidery: iron on top of a fluffy towel.
- Scorches: See Stains.
- To iron a shirt, iron the back first, then move to the arms, the front of the shirt and do the collars, cuffs and top of the front last.

- Don't wring clothes before drip-drying them.
- Remove clothes from the dryer before they are completely dry.
- Make a sharp crease by rubbing a piece of damp soap inside the crease. Then iron on the outside with a damp cloth. Or dampen crease with gum arabic. Or for dark fabrics, dampen a cloth with cold tea and iron over the crease.
- To get a crease out, soak creased garment in a solution of 5 cups (1.25 L) water with 5 tbsp (75 mL) vinegar for four hours, then dry until damp and steam iron.
- To make a garment crease-resistant, add one packet of gelatin dissolved into the final rinse.

J

JAMS AND JELLIES

- By substituting the juice of two lemons you can eliminate using commercial pectin and cut down the amount of sugar you'll need to use. Or cut up a quince and add to the jam instead of using commercial pectin.
- If you stew fruit, adding some fresh angelica stems to the mix will cut down enormously on the amount of sugar needed.
- Heat the sugar in the oven before you add it to the fruit, and add it when the fruit is tender and the juice is flowing. Then boil really fast for ten minutes.
- To figure out when jam is cooked, put a wooden spoon in it. When it falls off in two places, it's jelled.
- One quart (liter) makes 2 to 3 pints of jam.
- Wash, cut up fruit and add 3 cups (750 mL) sugar to 4 cups (1 L) fruit.
- Apple jelly: tart apples are high in pectin, so will jell more easily. Crab apples are the best. Add a rose geranium leaf to give an original bit of dash. Put it at the bottom of a jar and fill with apple jelly. A long needle or hairpin can be used to manipulate it if you want to suspend it in the jelly.

JAPANESE BEETLES

- Plant borage to trap these pests.
- Organic insecticides such as pyrethrum and rotenone must be squirted directly on them.
- Barriers of diatomaceous earth will help get rid of them.
- If you're lucky, starlings will attack the larvae in spring.

(*See also* Insect Control.)

JARS: Stuck Lids

- The easiest way to open a jar is to give it a good sharp bang flat against the floor. It will break the seal.
- Immerse just the lid in boiling water.
- To get a good grip, open the jar while wearing rubber gloves.
- Or wrap rubber bands, a damp sponge or dish towel around the reluctant lid.
- If dried food seals a lid, run warm water over the lid to soften food and expand the lid.

JEWELRY

- To store costume jewelry, put a piece of chalk in the box to keep jewelry from rusting or tarnishing.
- Use egg cartons to store precious items.
- Don't clean jewelry with toothpaste. Use a fine brush like an old baby's hairbrush and wash the jewelry in warm water and a mild pure soap. Gently rub it with the brush. Dry and polish with chamois.
- Clean diamonds by putting in boiling water and a little detergent and let stand for five minutes.
- Pearls will come clean if you wear them next to your skin.
- You can restring beads or pearls with dental floss: cut it 12 in (30 cm) longer than the old string and thread through with a needle.
- When you're restringing beads, lay them out in order on sticky or masking tape to prevent them from rolling around.

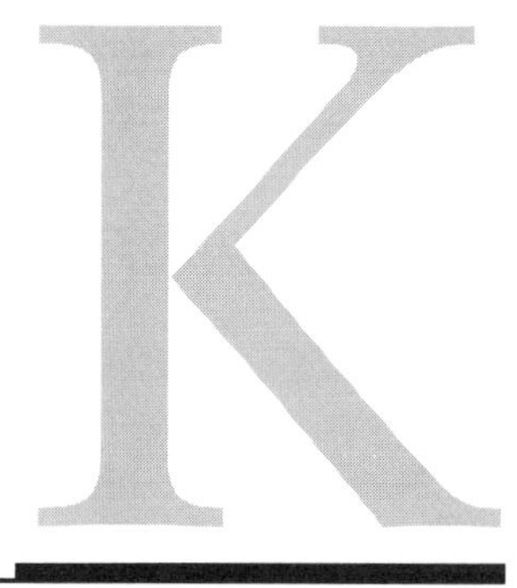

KEYS

- Keep a list of the code numbers of your auto keys—it will cost you less to have a new one made if you know the number.
- For deadlocks, you must not only have your secret number, you must return to the place where you purchased the lock to get a new key made.
- Keys locked in an older car can be retrieved by stretching a coat hanger to form a hook on the end. Jigger this down between the door and the car body. Pull the button up.

KITCHENS: CLEANING AND MAINTENANCE

You can save money and the environment by cleaning with home-made cleansers. If you prefer to buy cleaning products, choose environmentally friendly ones. (See Cleaning Formulas, Commercial Cleaning Products, Household Hazardous Products.)

Appliances (See Appliances: Cleaning.)

Cabinets:

- For hard to reach corners, install a lazy Susan.
- For deep corners, install a wide, V-shaped cabinet in the corner with a piano-hinge door. It will allow you to reach inside easily.
- To clean grease from wood cabinets, apply a very thin coat of car wax. Let dry and buff.
- To clean painted cabinets, try wiping with warm water first. For long-standing grime, mix equal parts of household ammonia and water. Be sure to wear rubber gloves, have good ventilation and avoid breathing fumes.

Floors:

- Remove crayon marks with toothpaste.
- Remove heel marks with an ordinary eraser. This is the safest way, of course, but if you have a really stubborn stain, considering rubbing it off with Varsol.
- To wax vinyl, wet the applicator first, wring it out, then add wax. You'll have more wax for the floor and less for the mop.
- Use an old detergent bottle to distribute the wax rather than sloshing it all over the place from the can.

Walls:

- Wall cleaner: mix ½ cup (125 mL) of household ammonia, ¼ cup (60 mL) vinegar and ¼ cup (60 mL) washing soda in 1 gallon (4 L) of warm water. Wear gloves to wash walls.
- Crayon marks can be removed by rubbing with a dry soap-filled steel wool pad. Or rub gently with baking soda on a damp cloth.
- Take off smudges with a clean art-gum eraser.

KNIVES

Always use a knife on a cutting board.

To Store Knives:

- The ideal storage is a wooden knife block or wall-mounted magnetic board. Use a divider to store knives in a drawer. This will prevent the blades from nicking one another.
- To keep carbon steel knives sharp, clean and rub with cooking oil before storing, or rub with a cork moistened in onion juice.

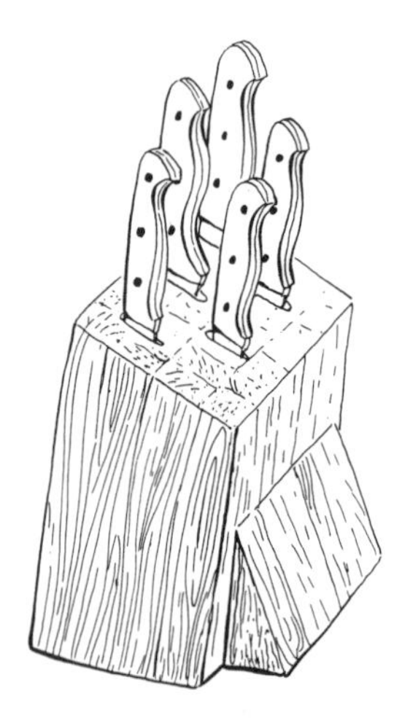

To Sharpen Knives:

- Use a bench stone if the knife is badly dulled. Apply light mineral oil to the stone before using. Some stones must soak in oil overnight. Hold knife at a 15° to 20° angle. Apply pressure with both hands. Draw blade in an arc over the stone. Reverse stroke. Repeat ten times.
- When using a chef's steel, always work on a 15° to 20° degree angle with light pressure. Draw knife toward you and to the right. Alternate strokes. About twenty strokes will do.

Cleaning and Repair:

- Any knives with hollow handles should not be put in the dishwasher. Hot water inside the handles could lead to rust or a weakening of the soldering that holds the handle and blade together.

- Clean bone-handled knives by rubbing a lemon dipped in salt over them. Rinse well.
- If the handle on a silver knife has come loose, reattach with epoxy putty.

L

LADDERS

- For roof work, have a ladder with hooks to attach to the edge of the roof.
- Make sure there's something to grab on to at the top in case you lose your balance.
- Attach antiskid strips to each rung on the ladder.
- To assure ladder safety, make sure there is at least one-quarter of the length in distance between the bottom of the ladder and the wall.
- On extension ladders, make sure there is at least 3 ft (1 m) of overlap.
- Put the feet of the ladder into empty coffee cans to steady them when the ladder is standing on soft earth.
- Glue weather stripping to the front of each step and you won't bark your shins.
- Don't wear sneakers. Wear hard-soled shoes with heels.

LAMPSHADES

- Never leave a shade in its plastic cover. Apart from looking tacky, the cellophane will shrink from the light bulb's heat and warp the frame. You'll end up with brown marks on the fabric.
- Avoid touching a lampshade with your hands—oils from your skin will discolor the cover and attract dirt.

Cleaning:

- To dry a wet lampshade, turn on the light, or use your hair dryer to do the job more quickly.
- To keep it clean much longer, apply a very thin starch solution with a sponge. (See Clothes Washing for a homemade starch recipe.)

- Vacuum with a soft brush attachment.
- Silk or rayon shades with fabric sewn rather than glued to the frame can be washed in mild, sudsy warm water and rinsed. Dry quickly with a hair dryer to avoid rusting.
- Rub spots on paper shades with a clean art-gum eraser.
- Clean parchment lampshades with a mixture of 2 tsp (10 mL) methylated spirits, 1 tsp (5 mL) phosphate-free liquid detergent, 2½ cups (625 mL) warm water. Mix well and apply with a damp sponge. Wipe off with clear water and let the shade dry.
- To keep parchment paper shades from drying out, rub once a year with castor oil.

LAUNDRY (See Clothes Washing.)

LAWN MOWER MAINTENANCE

The best favor you can do for your lawn is to get it off chemicals and start using a push mower. It cuts the grass in such a way that you'll end up with a better nap. Electric and gas mowers have the obvious problems—they pollute—and they pull rather than make a clean cut.

- Check blades to make sure they're on tightly.
- Rub the blades with vegetable oil to keep grass blades from sticking.
- Don't cut wet grass, or give grass that's too young a close shave.
- Never put the mower away without removing all the leaves and dirt from the machine. Wipe off any moisture on blades and handle to prevent rusting.

Gas and Electric Mowers:

- Every spring wipe off all surfaces, including those around the engine. Clean off the dirt from tires and wheel shafts.
- Start each year off with a fresh tank of gas and make sure you fill the oil tank. Always keep gas in a cool place to slow down oxidation.
- On the first outing of the season, run the mower for fifteen minutes, then turn it off. You'll find it runs better after that.
- Fill the gas tank after you've used the machine and watch for drips.
- Drain the oil and empty the gas tank at the end of the season. Disconnect the gas line from the tank before you drain it. Attach a hose to the wrong end of your vacuum cleaner and

blow it through the gas tank and fuel line. This will avoid oxidation during the winter.

- Avoid spilling gas. Even small spills can evaporate and pollute the air.

LAWNS

- To make a beautiful carpet of grass, your lawn will need about six hours of sun a day and 1 in (2.5 cm) of water a week. Otherwise, you might as well experiment with ground covers.
- To tell you how much moisture there is in the lawn, cut out a chunk 8 in (20 cm) deep and check the bottom. It should be moist. If not, water well.
- Water during the early morning, never at midday—you'll lose water in evaporation and run the risk of burning the grass.
- Fertilize with organic fertilizer, not chemicals. You'll kill off all the worms if you use chemicals and your soil will suffer terribly.
- Only put grass clippings that have not been treated with chemicals in the compost.
- Reseed in August and well into September.

LEAFHOPPERS

These nasty little beasts attack all sorts of vegetable crops. If you have a serious problem, try insecticidal soap, natural sprays and pyrethrum or rotenone.

- Do not plant beets or spinach near tomatoes since the beet leafhopper will attack the tomatoes and kill them.
- Dust plants with diatomaceous earth in late evening when beneficial insects are sleeping.

(*See also* Insect Control.)

LEAF-MINERS

- Cut up large rhubarb leaves and boil for three-quarters of an hour. Cool and strain, then spray on plants. If you use this on vegetables, be sure to wash them before eating—rhubarb leaves are poisonous.

(*See also* Insect Control.)

LEATHER

- When you buy any furniture with leather parts, spray them with water-repellant finishes just as you do new leather boots.
- Never use furniture polish, shellac or varnish on leather furniture.
- Clean leather furniture about once a month and don't polish too hard. Buff with a solution of two parts linseed oil and one part white vinegar. Let dry well.
- To clean leather furniture, use the mild soap and soft cloth, followed by an application of saddle soap.
- For stained leather-topped furniture, dampen a soft cloth with a mild soap and water concoction, then pat dry with a clean cloth and apply hard furniture wax.
- Water rings can be removed by sponging with white vinegar.

Leather Clothing:
Never use hot water on leather because it inevitably shrinks it, and never saturate leather with water.

- Wipe it off with a damp cloth and a bit of mild soap. Rinse with a damp sponge and let dry at room temperature. Rub gently with mink oil.
- Use saddle soap to keep clean. Rub it in with a damp sponge and when the leather is clean, dry off with a soft rag.
- To clean leather effectively, always test first in an inconspicuous area. Use a soft cloth and any of the following: ½ tsp (2.5 mL) borax dissolved in 2½ cups (625 mL) of warm water; eucalyptus oil (you can get it at the health food store).

Restoring Leather:

- After you've cleaned the surface, rub with a cloth dipped in raw linseed oil, castor oil or glycerine.
- Beat an egg white until it's stiff and rub into the leather surface. Polish with a soft cloth.

(*See also* Furniture Repairs; Luggage; Mildew; Shoes; Stain Removal: Fabrics; Suede.)

LEMONS

- When you use lemon in a marinade, be sure to use a glass bowl, not a metal container, which will be affected by the acidic quality of the citrus.

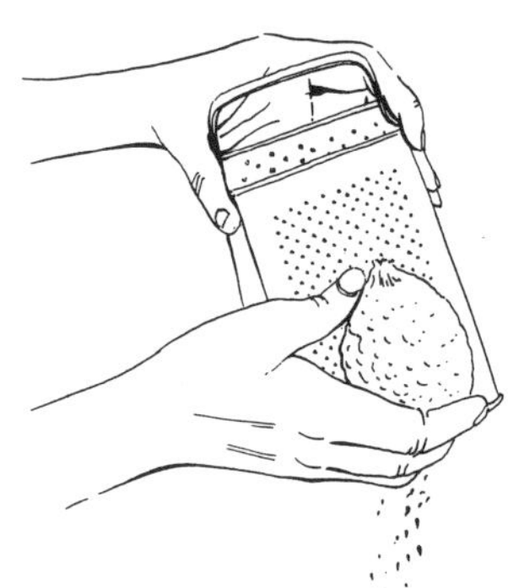

- Substitute ½ tsp (2.5 mL) vinegar for 1 tsp (5 mL) lemon.
- Heat a lemon in the oven or douse with boiling water and it will be much juicier.
- When you grate lemons, be sure to grate only the yellow part. The white part can be bitter. You can freeze grated lemon rind.
- A fast grating method is to peel the lemon and put the peel into the food processor.
- One tsp (5 mL) of lemon in rice water will make the rice much whiter. The same principle works on mushrooms if you are sautéeing them.

LIDS: Stuck (See Jars: Stuck Lid.)

LIGHTING

- Every two or three months it's a good idea to clean all your lighting equipment—bulbs, tubes, lampshades and fixtures. You'll be astonished at how dirty they can get and how much this cuts down their efficiency.
- Darkened light bulbs are inefficient. Place them in a dark closet or hall where less light is needed or which is seldom used.
- A tube with a dark end may mean it needs replacing.
- Delayed lighting may mean the starter needs replacing.
- Cleaning bases: glass, marble, chrome and pottery can be washed with a damp, soapy cloth and waxed after drying.
- Wood and metal bases can be dusted, followed by a thin coating of wax.
- Get into the habit of always turning the lights off when you leave a room.
- To clean a chandelier: take off all the removable parts and put them in the flatware basket of the dishwasher—this will make them sparkle once they're washed. Or wash by hand.

LOCKS (See Frozen Locks.)

LUGGAGE

- If a zipper won't work, rub a lead pencil along it.
- Use a mild detergent and sponge to remove soil from vinyl or plastic bags.
- Clean leather bags with saddle soap and a damp sponge.
- Don't stuff too much into luggage and never put any sharp objects along the sides.

(*See also* Packing Clothes.)

LUMBER BUYING AND STORAGE

- Beware of discount lumber—it could have something wrong with it—it may be green or warped.
- The lower the grade the more knots and defects for each board foot.
- Narrow boards have fewer defects than wider ones.
- When you order wood, you can try asking the lumber yard to stack the wood in the sequence you'll need it as you take it off the truck. They might just do it for you!
- If you have to wait a few days before using plywood, bundle the sheets together as you stack them up—this will keep them almost burglarproof. Don't nail them together, as you could injure yourself if you either forget the nails are there or if you don't get all of them out.
- Wood gradually shrinks as it dries. Kiln-dried wood is drier than air-dried wood; green wood has the highest moisture content of all.

MARBLE

Cleaning:

- To protect marble tables, use coasters under drinking glasses.
- If spills occur on any marble surface, wipe them up immediately.
- Clean tables weekly with a damp cloth, and floors with a damp mop.
- If marble becomes dirty or is streaked, wipe with a damp sponge and buff dry.
- Stubborn dirt: use dry borax and a damp cloth. Rinse well, then buff dry.
- Grease stains: treat with a mix of one part fine pumice powder, two parts washing soda, one part fine powdered chalk and one part white vinegar mixed with enough water to make a paste. Do a test in a small inconspicuous spot—the vinegar can etch marble. If you're happy with the results in the test spot, apply the solution, rubbing with a soft cloth. Clean with liquid detergent and water mixed with soda.
- For organic stains (coffee, fruit juices, etc.): mix powdered whiting or chalk dust with hydrogen peroxide and add a few drops of household ammonia just before applying.
- For rust stains: substitute liquid rust remover as the solvent. Do a test spot with this method, as well.
- If you drop ice cream on your marble, clean with the rust stain remover.
- A coffee stain takes the same treatment, except you add equal parts of hydrogen peroxide, household ammonia and cream of tartar, plus flour to make paste. Apply to stain, cover and leave for ten hours. Uncover and let dry. Brush off and wash with soap and water.

- If removing stains dulls the surface, wet the area, sprinkle with tin oxide powder and rub with a thick cloth or electric buffer. Always rinse well and then dry.
- Any minor imperfections can be sanded off. Use a very fine sandpaper soaked in water for about ten minutes. Wrap it around a sanding block and rub very gently along the grain. Keep the paper wet so that dust doesn't build up. Do not use this method on black or dark-colored marble.

Polishing:

- Clean the marble as suggested, then rub the marble in the direction of the grain with a soft cloth dipped in an automotive polish compound. Let dry and buff well. Finish with a light application of an acrylic polymer-based wax product, then polish.

MATTRESSES

An astonishing amount of hair, dead skin and other pollutants can get into your mattress, causing allergic reactions. Dust mites live there, too.

- Vacuum your mattress regularly. It will smell better too.
- Sprinkle mattress with a room deodorizer such as baking soda, then vacuum.
- You can also buy non-allergenic mattress covers, which can be washed.

MEALYBUGS

The cottony balls that invade your houseplants are mealybugs. Get rid of them this way:

- Dip a small plant into warm, soapy water and let sit for two hours, then rinse.
- Use denatured alcohol on a cotton swab and rub on the mealybugs, getting as little alcohol on the plant as possible. Then rinse thoroughly.
- If they get to the roots, wash the soil off with warm water and cut out the infected parts. Repot with sterilized soil.

MEAT

- Broiling: put a few slices of bread in the broiling pan under the rack to soak up drippings and lessen smoke.

- Red meat is medium rare when drops of pink show on the surface. If it's soft, it's rare; if it's hard the meat is well done.
- Sprinkle broiler with salt as soon as you've finished cooking and cover with wet towels. Or add a bit of sugar or a combination of cider vinegar and sugar to make it easier to wipe clean.

MENDING CLOTHES

The Kit:

- hand needles in all sizes
- pins (size 17)
- thread: black, brown (or any dark color to match your wardrobe), white and gray
- scissors
- glue stick to use instead of pinning or basting
- white candle stub to run thread across to keep from tangling
- extra buttons
- hooks, eyes and snaps
- tape measure
- magnet to pick up all the pins you drop
- small bar of soap to stick pins and needles in—keeps them in one place and lubricates at the same time.

Buttons:

- If you lose one, replace with a button from an out-of-sight part of the garment.
- With four-hole buttons, sew only two holes at a time, then break thread. They'll stay on longer.
- Use dental floss or elastic thread for buttons on clothes that get specially hard wearing.
- To remove a button, slide a comb under the button and cut off with a razor blade.
- To cut a buttonhole, lay fabric over a piece of soap and use a razor blade to cut.

Emergency Hems:

- Double-sided mending tape: place tape on the hemming line, but leave backing on tape. Press with fingers, remove backing and fold hem up at ends of the tape. Smooth hem by pressing with fingers from the center out. Remove before washing.
- Fabric glue: first test for staining. Spread foil under fabric and apply thin line to hem edge. Spread with finger. Fold hem in

place and blot excess with damp cloth. Dries in five minutes and can be washed but not dry-cleaned.
- Fusible web: cut to right length, apply strips between two layers of fabric and steam press. Can be washed and dry-cleaned.

Trouser Hems:
- Always try pants on with the appropriate shoes.
- Allow pants leg bottom to touch top of shoe in front and taper to slightly longer in the back. It should brush the shoe without wrinkling and dip in back.
- Measure second leg from the first.
- Clip hem in center ¼ in to ¾ in (6 mm to 20 mm) long. Spread fabric to fit inside pants leg.
- If a hem creases, sponge fabric with white vinegar and press. (*See also* Ironing.)
- When you're lengthening jeans or jean skirts, white hem lines will disappear if you mix a little permanent blue ink with water until you get the right shade and brush on. Let dry.
- If jean cuffs insist on turning up, iron patching.
- To keep pants from developing a hanger mark, cut a cardboard roll (toilet paper or paper towel) in half and put it over the hanger.

METAL STAINS
- Brass (See Brass.)
- Copper (See Copper; Pots and Pans.)
- Silver (See Silver Cleaning.)
- Stainless steel (See Pots and Pans.)

METER READING
Check your gas and hydro meters. Consumption is indicated by several little clock dials.

- Start with the right-hand dial and write down the number just passed. Move on to the next dial, which will run the opposite way and do the same. Keep the reading in a safe place and the next time you need to know your consumption, read the meter again. Subtract the lower number from the larger and you'll get the reading. Check with your local utilities office to find out how much you're paying for each unit (kilowatt hours in the case of electricity; cubic feet in the case of gas). You can calculate what your bill will be and check it against the utility's reading. Utility companies also provide the rate per kilowatt hour.

MICE: AS PESTS

Although mice are cute little creatures, they can be very destructive and rather messy.

- If you like cats and aren't allergic to them, consider getting one.
- Or, if a friend has a cat, see if you can have it stay with you for a few days. Mice will often find a new home.
- Mice will nest near food sources. Keep food in sturdy plastic or metal containers, as mice will chew through paper bags, cardboard boxes and thin plastic wrapping. Ensure garbage cans have a secure lid. Clean countertops, your stove and floor regularly.
- Find a humane trap. If you can, get a live trap and release the little creatures outdoors.
- Be sure to stop up all holes where they can get in. Steel wool is excellent, as mice hate to chew through it.
- When you stop up a hole make sure you aren't leaving bodies or live animals in there. Once they start decaying, you'll hate the smell and have to open up the hole again.
- In the garden, spread blood meal around and keep applying on a regular basis throughout the season, especially after it rains. Mice don't like the smell.
- There are poisons you can use, but I don't recommend them, particularly if you have children or pets.
- If you have an infestation of mice that isn't cured by the above measures, call a qualified exterminator.

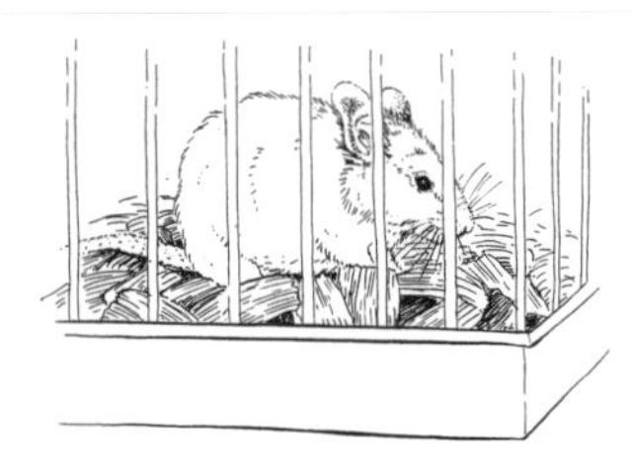

MICE: AS PETS

Choose a female as a pet—they are much less smelly—and one that is alert, sleek, active and curious.

- To pick up a mouse, cup your hand over it and slide your fingers under its body. It will bite if squeezed or frightened. If you are bitten, clean the wound with soap and water.
- To capture an escaped mouse, get a shoe box with one end open facing the mouse and herd the mouse inside. Tip up and return to its cage.
- Cage: The cage should be 2 cubic ft (60 cm^3). Put sand, sawdust or hay on the bottom, but not newspaper—the ink is lethal. Mice like to dig. They also choose an area in which to urinate and defecate.

- Teach it to walk from one hand to another. Let it climb on one hand and place the other in its path. Then reward it with a bit of food.
- Exercise wheels are important in avoiding boredom, especially during the night, since it is a nocturnal creature. Provide a nest box with bits of fabric and soft paper.
- Use toilet roll centers and branches to make a jungle gym.
- Your mouse will be happy as long as you play with it and talk to it daily.
- Feed it daily with mice pellets, dry dog food, seeds, nuts and bits of brown bread. Gather acorns in the fall and leave in the freezer until you need them. Variety is important. Change the water weekly.

MICROWAVE OVENS

Microwaves are more energy-efficient and timesaving than a conventional oven. However, all the extra packaging that goes with microwaving—disposable plastic trays, papers towels, foil and plastic wrap—are not environmentally friendly. If you have one, follow these rules:

- The amount of food in the oven increases cooking time exponentially: one potato takes four minutes, four potatoes take sixteen minutes.
- Use your microwave to thaw food, cook fresh or frozen vegetables and reheat food.
- Never use the microwave for cooking eggs in the shell.
- Do not use metal containers or containers with metal trim to cook food in a microwave: metal can cause an electrical arc that may damage the oven.
- Clean a microwave with a bit of baking soda on a soft damp cloth. Don't use anything abrasive.
- Cover a spill with a damp cloth. Turn on High for ten seconds and wipe clean once it's cool.
- Clean your microwave with 2 tbsp (30 mL) baking soda or lemon juice in a microwave dish of water. Let cook for five minutes. The steam will make it easy to wipe off the walls.
- Dip a rag in ashes from the fireplace and clean off oven glass. Wipe clean with newspaper.
- Rid your microwave of fish smells by putting a little vanilla in a bowl and turn on High for one minute. Repeat if the smell hasn't gone.

- Microwaves are ideal for melting chocolate or butter without burning it.
- Always put the dish in the center so it will cook evenly.
- Cover most vegetables and casseroles with a lid placed on an angle to allow hot gases to escape, rather than plastic wrap.
- You are going to lose a lot of nutrients by microwaving vegetables but some taste pretty good anyway. Artichokes are one of the few things that come out of this oven successfully. Pour lemon juice over the prepared artichoke, cover and microwave for fifteen minutes, depending on size.
- Don't believe anyone who says bacon is better in the microwave—it isn't. Cook it the usual way.
- Keep seals around the door clean. If they're cracked, replace immediately. You don't want radiation to leak.
- Keep at least a yard (meter) away from the oven when it's on.

MILDEW

General mildew remover: combine ½ cup (125 mL) white laundry starch, ½ cup (125 mL) lemon juice, ½ cup (125 mL) liquid soap and ¼ cup (60 g) salt. Apply to stain and leave in a well-ventilated place for two hours. Wash and rinse well.

- To avoid getting mildew in the laundry, put damp clothes in the freezer until you can iron.
- Keep a dehumidifier going all the time if you have a basement that's prone to mildew.
- Keep anything stored in the basement clean and store anything that's mildew-prone, such as books, somewhere else.

Books:

- Soak a cloth with 6 tsp (30 mL) copper sulphate dissolved in 2½ cups (625 mL) water. Let the cloth dry, then rub over books thoroughly. This will treat about fifty books.
- Sprinkle pages of a mildewed book with cornstarch or talcum powder.

Brick or Stone:

- Talk to a professional before undertaking cleaning brick—it's not really a do-it-yourself project. It's safe to hose down or pressure wash the brick down with water; you could also wash it with a non-ionic detergent, using a soft nylon brush.

Closets:

- Use mildew remover suggested above.
- Absorb excess moisture with small cloth bags of cornstarch, cornmeal, baking soda or talcum powder.
- Hang a small bag of calcium chloride in a damp closet and place in a bowl underneath to collect droplets.

Clothes:

- Brush and vacuum off as much as you can before you try anything else. Hang the items outside so the loose spores will blow away.
- Washables: for colored clothes soak in sour milk and let dry in the sun. Make sour milk by adding 1 tsp (5 mL) vinegar to 1 cup (250 mL) of milk.
- Colored clothes: don't use chlorine bleach; try soaking in a strong detergent solution.
- Whites: squeeze lemon juice on the stain and follow up with a bit of salt. Let dry in the sun, then wash.

Furniture:

- Brush fabric upholstery with a whisk broom, then sponge with a mixture of equal parts of rubbing alcohol and water, follow with a fungicidal spray.
- On furniture with painted surfaces, use 1 cup (250 mL) household ammonia, ½ cup (125 mL) vinegar and ¼ cup (60 g) baking soda per gallon (4 L) of water.

Leather:

- Leather books: take a cloth sprinkled with household ammonia and wipe any mold away. (*See also* Leather.)

Tents and Awnings:

- Moisten the mildewed spots with lemon juice. Sprinkle with salt and let dry in the sun.

MIRRORS

- If clips used to mount an unframed mirror on the wall are not padded, add thin adhesive-backed felt to prevent mirror from being scratched.
- To clean mirrors, spray with a mixture of one part white vinegar and five parts water.

- For badly soiled mirrors, wash with a warm solution of tea, water and phosphate-free detergent. Or use 2 tbsp (30 mL) vinegar, household ammonia or denatured alcohol in 1 quart (1 L) water.
- To disguise worn spots in the reflective backing, tape a piece of tinfoil to the back.
- To protect silvering, paint shellac on the back. This will extend the life of your mirror considerably.

MITES

In the Garden:

- A strong water spray from your hose is the best bet.
- Or try a spray of insecticidal soap mixed with light horticultural oil.
- A direct strike with pyrethrum, an environmentally-safe insecticide, will paralyse mites. Spray the undersides of leaves as well. Apply three days apart.

In the House:

If a plant's leaves turn yellow, test for mites. Hold a piece of paper under a discolored leaf and tap. You'll see them fall on the paper. Cut off the damaged parts and immerse the plant in lukewarm water.

- Stick the plant in the shower for twenty minutes—not too strong and not too warm.
- Mist your plants every day and don't let the atmosphere get too dry.

MORTAR

- Here is a general recipe for mortar: combine one part masonry cement and two and a quarter to three parts loose, damp mortar sand with enough water to make a plastic mix.
- For mortar cement, combine one part portland cement, one part hydrated lime and four and a half to six parts dry sand. This is a nineteenth-century method often used today on historic buildings.
- One cubic foot (30 cm^3) of masonry mortar will lay thirty bricks. One cubic foot of masonry mortar needs 31 lbs (14 kg) masonry cement, 100 lbs (45 kg) dry sand and 4 to 5 gallons (16 L to 20 L) water.

- You can add sand, gravel and aggregate stones to extend a cement-water batter.
- Dump the amount of mortar you need out of the big bag, put it in a plastic bag and add enough water to moisten. Knead the mix from the outside. You'll avoid dirty hands and equipment.

MOTHS

- Always wash clothes to be stored.
- Cedar does have some repellant qualities, but it's mild. Lavender is a pleasant-smelling retardant, but also mild. There are naphthalene mothballs and paradichlorobenzene crystals that smell terrible and are toxic. I don't recommend them.
- Place pieces of camphor, cedarwood, Russia leather, tobacco leaves, bog-myrtle or anything else strongly aromatic in drawers and boxes where clothing is stored.
- Store bay leaves in bags or boxes on shelves.
- Keep moths away with a few bars of toilet soap in storage boxes with woolens.
- Rosehips in a sachet will also help keep moths away.
- Hang cloves wrapped in cheesecloth with your clothes.
- Before storing clothes, sprinkling them with black pepper will keep moths out.

MULCHING

Mulch is the best weed killer there is. It also holds moisture in the soil, revitalizes flagging soil, feeds your plants and keeps them healthy. Never mulch with any material that's been sprayed with chemical herbicides or pesticides. Use grass clippings, pinecones, cocoa bean hulls, chopped up leaves, compost (the best mulch in the world), seaweed, straw or hay (without any seeds), shredded bark or stones.

- Mulch in the fall after the ground has frozen.
- Remove mulch in spring during your cleanup around the garden and put it in the compost pile. Then refurbish the mulch.
- Don't put mulch right up against the delicate stems of plants.

NAILING WOOD

To prevent wood from splitting when you use nails:

- Blunt the tip of the nail by filing it or hitting it with a hammer.
- Drill a small hole in hardwood before you start hammering the nail in place.
- Don't drive the nail into the same line as the grain of the wood. Stagger nails rather than placing them along the grain line. For extra holding powder, use coated nails or nails with spiral shanks or annular rings. On outdoor projects, use galvanized nails, which won't rust.
- When you are using either screws or nails in very hard wood, use ordinary wax—paraffin—or graphite as a lubricant. Don't use soap since it will have a moisture content that could cause rust or rot.

NAILS: POPPED OUT

- If a nail partially pops out of its place, replace it with a larger nail.
- Or drive another nail 1½ in (4 cm) above or below it. Then drive in the popped nail.
- If you're hammering nails into wood, hammer the nail until it's close to the wood, then finish the job with a nail set to countersink the nail. Don't use nails on drywall—use screws.
- Fill any dents with joint compound. Sand before painting.

NEMATODES

These are nasty little worms that eat plants from the root up.

- Add compost with kelp meal as a mulch. The fungi that develops will attack nematodes.
- Fish emulsion repels nematodes.
- Castor beans are a good decoy crop.

NUTS

- If shells get mixed with meats, put the whole lot in a bowl of water. Shells will float, so scoop them out.

ODORS

- If the inside of an old box or trunk smells, moisten a piece of bread with vinegar and leave inside overnight.
- Put a lump of charcoal in a container before storing.
- Get the smell out of plastic containers by leaving a piece of crumpled newspaper inside for a few days.
- Remove odors from shoes and boots the same way.
- Or sprinkle baking soda in shoes and boots.
- Get smells out of your skin by washing in a solution of vinegar and water. Or use a slice of cucumber or just the peel and rub hands fresh.
- For getting rid of musty odors in suitcases, soak cotton balls in vanilla extract and leave overnight. Or try kitty litter and leave it closed for a few weeks, then vacuum.
- To get the musty smell out of a mattress, sprinkle baking soda on top and bottom. Cover with a sheet and leave for a week. Vacuum before using.
- Fridge odors can be eliminated by washing thoroughly with baking soda and water. Place a box of baking soda in the fridge to absorb odors. Remember to replace it every few months. A piece of garden charcoal will also absorb odors. It can be reused by heating gently in a heavy pot.
- If you have a mothball odor inside a piece of old furniture, sand down the inside and then apply a coat of urethane.
- Furnace oil smell: call a serviceman.
- Perfume odor: if it gets into furniture, leave garden charcoal in a bowl in or near it for a few weeks.
- Clothes: soak for thirty minutes in equal parts of vinegar and water.

- If books smell funny, put them in a bag with biodegradable clean kitty litter. Seal tightly and let rest for a week.

Cooking Odors:

- A cup of vinegar on top of the stove will cut back odors of smelly vegetables, such as cabbage.
- Add sugar and cinnamon to a pot of water and simmer gently. It will mask just about anything.
- An even more effective brew is to put cinnamon sticks, potpourri that is about to fade away, or a bit of lavender in a pot of water on top of the stove when you are cooking. It will scent the whole house. A bit of borax will make it even better.
- Pomanders (oranges, lemons or apples stuck all over with cloves) will last up to seven years. Hang in the bathroom and you won't have any offensive odors in there.

Room Odors:

- Start by airing the room out.
- Rather than using commercial sprays, place containers with a few drops of vanilla extract in them in different parts of a smelly room to purge it of odors.
- Boil ten cloves in 1½ cups (375 mL) water to clear the air.
- Leave vinegar in bowls around to absorb odors.

(*See also* Animal Odors; Cigarette Odors; Refrigerators.)

OLIVE OIL

- Olive oil will last for two years if kept in a cool dry place, but it is at its best and should be consumed within the first year.
- Keep a small amount of oil in a spray bottle by the stove. Spray pans, fish and fowl—this is a lot easier than brushing on.
- You can substitute oil for the fat used in baking: ½ cup (125 mL) shortening equals ½ cup (125 mL) oil.
- Combine a little butter in oil for sautéeing and it won't burn.

ONIONS

- To sweeten raw onions, cut onions in very fine rings. Put in a bowl of ice-cold water. Squeeze the rings then let go to extract the acid. Do this three or four times, changing the water each time. You'll have sweet mild onions to eat raw.
- Once an onion is cut, rub it with butter before putting it in the fridge.

- Sprinkle onions with a bit of sugar and they will sauté evenly.

ORANGES

- Pour boiling water over oranges, let stand for five minutes and they'll be a snap to peel. You can even refrigerate after and the peel will be permanently loosened.

OVENS

(See Stoves and Ovens.)

PACKING CLOTHES

Leave this travel care package in your suitcase all the time:

- a bag with cold-water/gentle soap
- one large and one finer needle
- several samples of thread wrapped around a piece of cardboard
- glue for emergencies
- safety pins
- folding scissors
- a small selection of ordinary buttons

Some general packing tips:

- Stuff socks and stockings in shoes and wrap in socks.
- Roll lingerie together and pack a bag to be used as a dirty laundry bag for the return.
- Don't roll belts; they can crack.
- Always pack cosmetics tightly in cosmetic bags.
- Fold clothes so that they interlock: pants first, followed by jackets, coats and skirts. Layer with tissue. Fold into the center. Blouses and shirts: fold sleeves to back and alternate direction. Fold ends into the case.
- Place small, rolled items into corners.

PAINT DISPOSAL

Paint should never be disposed of by pouring it down the sink or throwing it out in your local garbage pickup.

- To prevent having leftover paint, buy as much as you need for the job.

- If you do have leftover paint, consider donating it. High schools and local theaters could use it when they build and decorate sets.
- If you must dispose of it, call your community's hazardous waste disposal facility.

(*See also* Household Hazardous Products.)

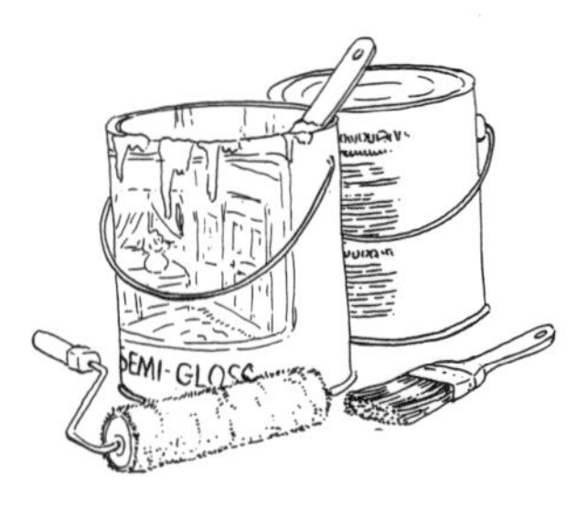

PAINTING EQUIPMENT

Brushes:

- An 8-in (20-cm) brush covers the largest area but is hard on the wrists, a 4-in (10-cm) is easy to handle but will take longer, and a 2-in (5-cm) is excellent for edges and corners.
- A brush will last longer if you soak it in a can of linseed oil for about a week before you use it. Clean it well before you use it, as any remaining oil will keep the paint from drying.
- Never overload the brush. Dip the first third of the brush into the paint, but no more.
- Apply paint in vertical strokes, spread at right angles to make an even coverage.

Pads:

- Use an 8-in (20-cm) pad for walls and ceilings.
- Draw the pad across the tray so paint is picked up evenly. Keep it firmly on the wall and sweep evenly in all directions. For oil paints, finish with vertical strokes.

Rollers:

- Use a 9-in (23-cm) roller for ceiling and walls, a long nap for textured surfaces and with latex, add a short nap for smooth surfaces and oil paints.
- Rollers with long extension handles for ceilings will eliminate the need for ladders.
- Always apply oil paint in one direction—usually toward the light.

To Clean Brushes and Rollers:

- Latex: wash in warm soapy water, rinse until water is clear. Run a comb or fork through bristles. If you're going to be using the brushes the next day, leave the soap in the brush; it stops the bristles from splaying out as the soap acts to hold them together. Rinse the soap out just before you use them.

- Oil: swish through a small amount of paint thinner or turpentine. Repeat until there is no more color on the equipment. On last cleaning, use soapy water. Hang to dry. Decant the solvent and reuse.
- Soften hard bristles with hot vinegar then wash with warm soapy water. Hang to dry.
- To clean paintbrushes, put in a plastic bag and pour in brush cleaner. Close bag and work the cleaner into the brush. You'll use less solvent this way.
- An empty milk carton is a good container to soak roller pads in. Just cut the top off and put the roller inside.
- When you clean used paintbrushes, use cans with covers. Let paint settle for a couple of days and pour out paint thinner for reuse.

To Store Brushes:

- Dry after cleaning, then wrap in foil to help retain its shape and hang.
- If you'll be using brushes again soon, wrap in plastic and store in the freezer.
- Or drill a hole in the handles, attach a piece of wire from a coat hanger and suspend the brushes in water (for latex) or paint thinner (for oil).

PAINTING: Inside

- Always buy the best quality brushes and paint you can.
- The paint: latex tends to be difficult to handle in hot weather; oil-based paint tends to be difficult to handle in colder weather. The older the paint and the cheaper the quality, the more likely it is to contain mercury and the harder it will be to handle. The most environmentally-correct paint is water-based latex. (Aura is a good product.) You can use it in most situations and the eggshell variety is relatively easy to wipe clean. Oil-based paint gives off toxic fumes. If you must use it, always paint in a well-ventilated space.
- The brushes: cheap brushes have silky bristles that tend to break easily. Paintbrush bristles should be springy and elastic. Hit the brush against your hand or fan bristles to see if they come out easily. A good brush will get better with use.

Walls:

- Paints can vary in the coverage you'll get. Generally, the more you pay, the further the paint will go. As a rule of thumb, plan on one gallon (4 L) of paint for every 450 square feet (40.5 m^2) of wall.
- Always paint the ceiling first, followed by walls and then trim.
- Coat water-stained areas with an alcohol-based or alkyd stain killer before painting.

Ceilings:

- Loosen the canopy on the chandelier or other lighting fixtures (after you've turned off the power) and enclose fixture in a bag.
- If you use a brush, shove the handle through a paper pie plate to cut down on paint falling on your head or running down your arm.
- Wear goggles to protect your eyes.
- Start in a corner near a window with a small brush. Paint a strip around the edge of the room before painting the ceiling.

Windows:

- Before you start painting, be sure to remove all dirt with an old paintbrush. Start early in the day so you can close windows at night without them sticking.
- If you have to reputty the window, mix the window color in with the putty—even if it's white.

To keep windows free of paint try the following:

- Spread damp strips of newspaper around the interior edges of the window. They'll cling until they dry—by that time you'll be finished painting.
- To get drips and splatters off windows, use a paint scraper that has a razorlike blade. Or soften the old stains with turpentine and scrape with a razor blade.
- Remove fresh paint with a hot vinegar solution.
- When painting windows, coat lemon oil on woodwork and splatters will wipe off easily.

Woodwork:

- An angled sash brush makes it easier to do hard-to-reach trim.
- Be sure to work on a clean surface, sanded smooth.
- Start with doors, frames and windows and end with baseboards—any dirt accidentally missed won't be moved around.
- Don't go over paint that's started to dry—it will leave a defect.

PAINTING: Outside

- Choose a moderately cool time in spring or fall.
- Follow the shade around the house, you don't want direct rays on the job.
- Wait two days after it rains before beginning.
- Use flat paint for siding. Latex is the easiest. Use gloss or semi-gloss for trim.

Primers:

Again, use organic primers. They are especially good on new bare wood. You can also find organic antirust primer for metal surfaces. Check with your local environmental store. New products are coming out all the time. Always looking for an alternative. If you can't find an alternative, use a primer put out by the same manufacturer as the paint you'll be using, or use the following:

- Oil-based primer for new bare wood, preprimed wood and siding; old wood surface; heavily chalked surfaces.
- Antirust primer for metal surface.
- Galvanized-metal paint for bare galvanized metal.

Siding:

- Repair any damage to siding; repair caulking and pound in any protruding nails.
- Wipe new metal with mineral spirits to remove the manufacturer's protective coating.
- If aluminum siding has an anodized or baked-enamel finish, be sure to sand it before priming.
- Wash any dirty areas with phosphate-free detergent and a stiff brush. Hose down and let dry. Or use a pressure washer. Cleaning is the most important thing you can do with siding.
- Nail-head staining: Countersink nail below the surface of the siding and spot-prime with exterior wood primer. Fill with caulk and apply two coats of paint.

(*See also* Aluminum Siding.)

Wood:

- Mildew: sponge off with a solution of 1 tsp (5 mL) phosphate-free, nonammoniated household detergent, 1 quart (1 L) household bleach and warm water to make 1 gallon (4 L). Wear gloves and goggles.

- Flaking, cracking and blistering: scrape off loose paint and prime.
- Chalking: scrub with stiff brush and detergent. Rinse and prime with oil-based primer.
- Fill in cracks with exterior Spackle, then scrape and sand patched area.
- If there are recessed nails, putty and sand before priming.
- Use a quality primer and make sure it's compatible with the type you'll use as the finish.
- Cover shrubs and other plants as you work.
- Start at the top, work down and paint an area 3 to 4 ft (.9 to 1.2 m) square at a time.

Trim:

- To protect a paint job on the siding, wrap rags around the ends of the ladder. (*See also* Ladders.)
- To get the best finish, sand to remove anything loose and feather edges of old paint. Cover the surface with primer, dry and sand lightly to smooth before painting.
- Gutters and spouts: after cleaning, prime bare spots with metal primer and coat inside with asphalt paint. If using latex top coat, apply directly on bare spots after cleaning with paint thinner.

PAINT PROBLEMS

- Don't apply a second coat before the first has dried or you'll end up with "alligatoring"—a wall that looks like it sounds.
- Don't apply paint in hot, direct sunlight or it will blister.
- Don't apply paint over a greasy, dirty or damp surface or it will peel.
- Don't apply a paint too thickly or it will wrinkle when it dries.
- If dirt or fluff gets into your paint job, let it dry, then sand and touch up.

PAINT REMOVAL

Liquid paint remover is incredibly toxic; use a machine instead. If you are removing old paint, it may contain lead. Wear a good-quality face mask and old clothes. Dispose of the mask and clothes safely when you've finished. Never remove paint in a place that isn't well ventilated.

- Rent an electric paint remover to take thickly-layered paint off a board surface. The higher the wattage, the faster it works.

- Soften paint splatters on a window with hot vinegar, then scrape off with a razor blade.
- Don't let paint stay on your skin. Wash it off each time you stop with soapy water or baby, mineral or cooking oil.
- Get paint out of a carpet by sponging with a solution of 1 tbsp (15 mL) phosphate-free detergent and 1 tbsp (15 mL) vinegar in 1 quart (1 L) of warm water.
- Make a paste of starch and hot water and apply to spills on vinyl floors or linoleum. Let sit for thirty minutes. Wipe clean.
- If paint smell gets to you, put a couple of chopped onions in a pail of cold water in the center of the room. One will cut down on the smell of the other. If that doesn't work, put a halved onion or a cut lemon in a dish.

(*See also* Stripping Furniture and Woodwork.)

PAINT STORAGE AND REUSE

- Eliminate drippy cans by driving nail holes along the edge so the paint can drip inside rather than outside.
- Bang lids on tightly and store upside down so scum won't form.
- Placing a circle of aluminum foil on paint surface the same size as the can will also prevent scum.
- Store paint in an opaque plastic milk jug and you'll be able to tell what's inside. Or label accurately, and attach a paint chip to the can.
- Mark the level of paint on the outside of the bucket. Oil-based paints can be kept fresh with a thin layer of 4 tbsp (60 mL) mineral spirits on top. Mix next time you use.
- Strain old paint through an old pair of discarded panty hose.
- Keep lumps at bay by cutting a piece of screen the size of the lid and let it settle into the paint. Lumps go to the bottom.
- Always keep a paint sample in a cool dark place in case you need a remix. Use popsicle sticks or lids from plastic containers.

PANELING (See Wood Paneling.)

PANTS FITTING

- Put the pants on right side out and pin to fit body.
- Remove pants and mark new seams on the inside. Remove pins and baste. Check fit.
- Across the derrière: let out back-seam allowances on side seams. Raise seam at center back. If too tight, lower the crotch.

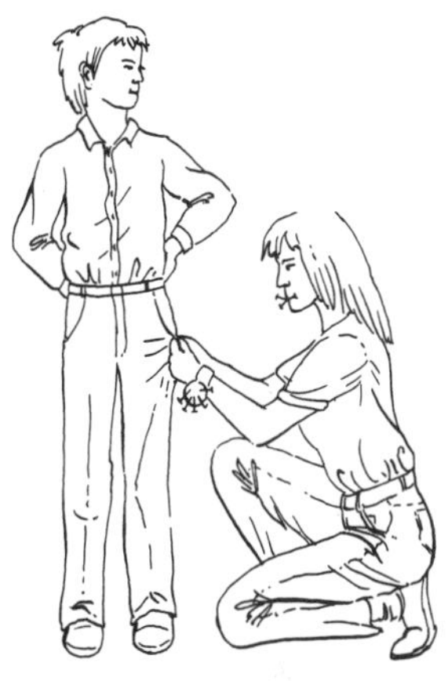

- Sagging across the derrière: make a deeper waist seam in the back. Or take in back side-seam allowances.
- Wrinkling in the crotch: make a deeper crotch. Lower a bit at a time until they no longer wrinkle.
- Tight or loose legs: add or subtract up to ⅜ in (1 cm) at the side and inner seams. (*See also* Mending Clothes: Trouser Hems.)

PANTY HOSE

- Double the life of your panty hose by saving them: cut off the damaged leg and match it up with another pair with a good leg in the same shade. You'll have a double top.
- To make new panty hose as runproof as possible, dampen and place them in a plastic bag in the freezer. Thaw and hang up to dry.
- Put panty hose and other small items in a mesh bag (the kind onions come in) and they won't get tangled in the laundry.
- Use old panty hose as a brush on suede.
- Use old panty hose to tie plants to supports.

PASTA

- Pasta always cooks better with 1 tbsp (15 mL) olive oil in the water.
- Fresh pasta is done al dente when it has come to a boil again for a minute or two and is floating on the surface of the water.

PATCHING CLOTHES

- Patches are effective on holes you can't darn.
- Find a piece of material from somewhere on the garment where its removal won't be noticed. If it's patterned, make sure the pattern matches.
- Make a rectangle about 1 in (2.5 cm) larger than the hole. Put patch under hole and pin. Stitch with machine or by hand. Clip corners to the stitches. Turn edges under and press.
- Slip stitch folded edges. Then catch stitch edges of the patch to the inside of the garment.

If you are rushed for time, try patching this way:

- Buy a precut, iron-on patch.
- Use mending fabric or tape and fusible web.
- Apply the patch with fabric glue or adhesive.
- Machine stitch a patch with a decorative stitch pattern.

Elbows and Knees:
- Make a paper pattern and cut out fabric or fusible web from it.
- Or buy a decorative applique patch of synthetic or genuine suede. Baste or fuse in place by putting fusible web between garment and patch.
- Put brown paper under worn area to keep fusible web from sticking to the inside of the garment.

(*See also* Darning; Mending Clothes.)

PEARLS (See Jewelry.)

PESTICIDES: Household (See Insect Control.)

PESTICIDES: On Produce
There has been increasing worry about pesticides on store-bought vegetables and fruits. Cutting produce out of your diet is not a suitable solution, as the beneficial effects of fruits and vegetables on the body outweigh pesticide risks. To reduce your exposure to pesticides, try the following:

- Buy organic foods whenever you can. And support stores and growers that offer organic foods.
- Rinse all fresh fruits and vegetables thoroughly with water.
- Fruits and vegetables with edible peels should be scrubbed with a brush.
- Remove outer leaves of lettuce and other greens.
- Peeling helps rid produce of pesticides, but you will often lose fiber and nutrients by doing so.
- Avoid washing produce with detergent. Although it is effective at removing pesticides, it can leave a residue of material that isn't fit for consumption.

PET GROOMING
Cats:
- Many cats can come to love being groomed with the vacuum. Start the habit by giving your pet some catnip and follow it with vacuuming.
- For long-hair cats, use a currycomb. It does a more effective job than a brush.
- Short-hair cats will need shampooing at least once a year; long-hair cats will need shampooing twice a year.

Dogs:

- Clean your dog's coat by rubbing oatmeal into it and brushing it out. It will bring any dirt and grease with it.
- Brush baking soda through his coat to deodorize your dog.
- Train your dog to enjoy a weekly brushing by talking to him, rewarding him with hugs.
- Don't bathe a dog that has dandruff—it will make things worse. Add 1 to 3 tsp (5 to 15 mL) corn oil a day to its regular diet.
- Add two raw eggs to regular diet each week.
- Remove stains from a white coat by a solution of 3 percent hydrogen peroxide and water. Blot the areas with this, followed by a brushing of cornstarch. Brush when dry and keep on doing this each day until stains disappear. Do not do this around a dog's eyes.
- Make a powder of equal amounts of fuller's earth and cornstarch and put it in a squeeze bottle. Squirt this on stain around whiskers and brush out.
- Condition your dog to the vacuum cleaner and you'll be able to clean up molting fur quickly.
- Always groom in the same area, use soothing words and make it a pleasant experience for both of you.
- Following the grooming tips above means you'll only have to shampoo your dog once or twice a year.

Bathing Cats and Dogs:

- Make sure the animal has relieved itself first.
- Get it well brushed and in a good mood.
- Make a ring of flea-soap lather and rub it around the neck to stop fleas from marching forward.
- Be sure to rinse very thoroughly—they'll inevitably lick themselves.
- Make your pet's coat shine with a rinse of lemon or vinegar and a little baking soda added.
- After the shampoo, move the dog to the shower stall and let it have a good shake, then towel dry.
- Keep any animal warm and dry after shampooing.
- The little beasts will want to roll around in the nearest dirt after you've gone to all this trouble—get out the vacuum cleaner.
- Putting a thick towel in the bottom of the sink will give the cat something to sink its claws in.

(*See also* Animal Odors.)

PEWTER

- Never store pewter in oak drawers or cupboards. Oak gives off corrosive acids that will damage pewter.
- If pewter corrodes to the point of becoming spotted or covered with gray scale, remove with equal amounts of hydrochloric acid and water. This is dangerous stuff and should only be used in desperation and with caution. Always wear rubber gloves and work in the open air.

PIANO KEY CLEANING

- Keep ivory keys uncovered—they need the light to keep from yellowing.
- Take a damp cloth and pick up bit of baking soda on it, then gently rub each key. Wipe clean with a damp cloth and gently buff dry.
- To whiten ivory keys, rub with equal parts of lemon juice and water, but don't get any on metal parts. Wipe with a damp cloth and buff dry.
- Mix ½ cup (125 mL) vinegar to 2 cups (500 mL) water and clean keys with leather chamois. Wipe dry and leave keyboard open.
- Make a paste of lemon juice with a little whiting and rub on the keys. Let sit for an hour and wipe off with a little hydrogen peroxide.
- Drip a cloth in eau-de-cologne, then in talcum powder (or powdered milk) and rub over keys.

PICKLING

- Use any fresh fruit or vegetable within twenty-four hours of harvesting. Wash thoroughly.
- Use pickling salt with no additives and distilled white vinegar with 4 to 6 percent acidity.
- Use only fresh whole spices; old ones will make a dark brine.
- For crisper pickles, add unsprayed grape or cherry leaves during the brining, then discard. Don't use lime or alum for crisping—you'll end up with mushy pickles.
- Use glass jars for long-term storage.
- If you use garlic, blanch it for two minutes before adding to other ingredients.
- Store in a cool dark place and keep an eye on the jars. A bulging lid may mean leakage or spoiling—get rid of the jar immediately.

PIPE INSULATION

- Cut down heat loss by insulating hot-water and heating pipes.
- Insulate cold-water pipes to reduce condensation.
- To measure for the correct-sized sleeve of insulation, close an adjustable wrench over the pipe and measure the distance between the jaws.

PIPES

- Plumbing pipe is sized by its inside diameter; a ¾ in (2 cm) pipe has a ¾ in (2 cm) opening, no matter what it's made of or how large it is outside.
- To repair a leaky pipe, an epoxy compound may do the job if the pipe is PVC.

(*See also* Frozen Pipes; Plumbing.)

PLANTS (See Garden; Houseplants.)

PLASTERBOARD OR DRYWALL

- Use water-resistant sheets of plasterboard or drywall for a kitchen or bathroom.
- Do not put a joint between two sheets along a window or a door edge; settling and frame shrinkage will cause cracks.
- Make cutouts for windows and doors as you go. For accuracy, measure from the ceiling down.
- Plasterboard patches: Dig out loose gypsum and fill with spackling compound.

PLASTER PATCHING

- Run a screwdriver along a crack to loosen particles, leaving a V-shape. Brush out.
- Wet edges with water and apply a plaster-bonding agent.
- Working from the edges to the center, fill the crack to half its depth with patching plaster or spackling compound.
- Finish patch by stroking with fine sandpaper.

PLAY SAFETY

For Babies:

- Never give babies glossy magazines. They'll suck on them and the ink sometimes contains lead.
- Never give infants packaging material. It can be swallowed, inhaled or shoved in an ear or nostril.

- Never give a baby or a small child anything sharp, with long strings or ribbons, or small things that can fit easily into the mouth.

For Older Children:
Give any playground an inspection before you let your child use it. Make sure of the following:

- Swings should have restraining devices so children won't fall off.
- Seesaws should have enclosed moving parts so that little fingers don't get caught.
- The merry-go-round shouldn't have a spot between the moving platform and the ground where a child's foot could get lodged.
- Slides should be built over sand or grass, not concrete.
- Cover backyard sandboxes when not in use to prevent them from becoming a kitty-litter box for stray cats.

Putting Toys Away
- Always put toys away so they don't constitute a hazard for anyone in the family. Use lightweight baskets or plastic boxes—stacking boxes are the best.

(*See also* Childproofing Your Home; Internet Safety.)

PLUGS: Overheating
- If an appliance heats up at the plug, the problem may be a connection in the wall receptacle. If the cord overheats, the appliance probably needs repairs. Check to see if the spring clips inside the wall receptacle have loosened. Be very careful working with anything electrical. Don't forget to turn the power off.

PLUMBING
Types of material used:
- Plastic is easy to identify. PVC is always black, ABS is always gray.
- Copper is the color of a penny.
- Brass: scratch the surface to find a yellowish color.
- Galvanized steel: apply a magnet—it will stick to steel but not to brass.

If you have trouble with a washer, try the following:

- If the washer-retaining screw is jammed on, use a penetrating oil to loosen it. Or use an appropriate wrench or a screwdriver with a bigger handle to give more power.
- If the screw slot is damaged, create a new slot by a few quick strokes with a hacksaw across the top.
- Try tapping the screw to loosen it with a hammer.
- Cola also works.

If the head of a washer screw breaks off:

- Expose the screw by digging out the surrounding washer with a penknife.
- Scrub most of the rust off with a wire brush, then give it a jolt by gripping the connection with a pipe wrench and smacking the wrench with a rubber mallet.

PLUMBING FIXTURE STAINS

- Stains from dripping faucet: wash with soapy water to which you have added ammonia. Soak rag with this. Place over stain and leave for several hours.
- White deposit around faucets: soak rags in vinegar and leave on stains for two hours.
- Vinegar stain on vinyl tile: use automobile rubbing compound or #0000 steel wool to rub off the stain, then polish and buff as usual.
- Use the juice from half a grapefruit and 1 tbsp (15 mL) rust remover on stains around sinks and tubs.

POISONING

- Try to find out precisely what was taken. Read the label for instructions and the antidote.
- Don't try to make the patient sick or give any salty drinks, especially if what's been ingested is corrosive.
- If you have small children, it is wise to keep ipecac syrup (bought at the drugstore) in the house. Keep these single-dose containers in a high, safe place. It is followed by several glasses of water or juice to bring on vomiting if it's necessary.
- Go immediately to the nearest poison-control center and take a sample of the poison with you.

POISONOUS PLANTS

The following plants are some of the common poisonous plants you shouldn't have around when you have a crawling child or curious pets:

- Indoor plants: philodendron, English ivy, hyacinth, daffodil, narcissus, poinsettia, azalea, hydrangea, diefenbachia, caladium, elephant's ears, mistletoe Monkshood, False Indigo, Autumn Crocus.
- Outdoor plants: aconitum, baptisia, colchicum, Lily-of-the-Valley, convallaria, crinum, daphne, euphorbia, iris, kalmia, azalea, laurel, rhododendron, rhubarb, ricinus (castor bean), sweet pea; daphne berries, jasmine berries, privet, yew, jimson-weed, deadly nightshade, wisteria.

POTATOES

- To keep potatoes from sprouting, add an apple to the bag.
- If you add 1 tsp (5 mL) lemon juice to the cooking water, peeled potatoes will stay white.
- To make old potatoes new, soak them in cold water and keep in the fridge for a few days.
- Recycle the cardboard cups that apples sometimes come in by storing potatoes in them in layers. This will give them a much longer life.

POTS AND PANS

With proper care, your pots and pans will last a long time. Some cleaning tips:

- A general rule to clean burned pots that still have food clinging to them is to add some vinegar to water, bring to a boil, then take the pot off the heat and let stand for at least an hour. Wash and dry as usual.
- Soak burns with undiluted vinegar. It will lift off the burned stuff if you can afford to leave it long enough.
- Add lard or shortening and vinegar to cover the bottom. Let stand for a while, then clean as usual.
- Apply 1 tbsp (15 mL) cream of tartar, add water and boil for ten minutes.
- Baking soda makes pots clean and shiny.
- Boil unpeeled rhubarb without leaves for twenty minutes in a scorched pot to brighten it up.

Aluminum Pots:

- Keep shiny by cleaning with lemon juice and mineral water.
- There is still a persistent belief that it is dangerous to cook with aluminum pots. This was due to research from the 1970s that proposed a connection between Alzheimer's disease and aluminum levels in the body. However, other researchers have not duplicated the findings.

Cast Iron:

If you prefer the even heat that these pots provide, as well as their ability to increase the iron content of food, keep the surface properly maintained and they'll last longer than nonstick pans. Keep these tips in mind:

- Never scour cast iron with soap, wipe clean with paper towel or a damp cloth.
- Use a nontoxic oven cleaner on the outside (*See* Stoves and Ovens: Cleaning). Let stand for a few hours and remove with vinegar and water. Don't do this on the inside. To keep the cast iron shiny, rub lightly with mineral oil.
- Rust is the big enemy here, so store your cast-iron pots and pans in a dry place and never use them to store food in—any moisture will cause rust. You need iron in your diet, but not necessarily this way.
- Season cast iron by washing the pan with a plastic scrub brush and mild, soapy water. Then coat the inside with salt-free cooking oil (not olive oil however, it may produce an after-taste or odor). Roast the pan in an oven set at 350°F (176°C) for two hours. When it cools, wipe off an excess oil. Then rinse the pot in clear water and dry by allowing it to sit on a burner set at low heat for a couple of minutes.
- A less complicated method for pots with wooden handles: spread a light film of cooking oil over the bottom and put on heat until it begins to smoke. Remove after ten minutes and cool. Wipe off any leftover oil.
- Each time you use a cast-iron pot, wipe it clean with a damp cloth, then dry it thoroughly.
- Remove burned food by scraping with coarse sea salt. Or start all over again and season.

Copper:

- Copper-bottom pots: apply ketchup and spread on the bottom. Leave five minutes and rinse with soap and warm water.
- Rub with half a lemon dipped in salt, rinse and polish.
- Mix up flour, salt and vinegar in equal parts. Smear this paste over the tarnished copper. Wash, then polish.
- Or spray it with a combination of 10 oz (300 mL) vinegar and 3 tbsp (45 mL) salt. Allow to set, then gently scrub clean. Rub with lemon halves dipped in salt or with Worcestershire sauce to get rid of tarnish.
- It may seem obvious, but do not cook with copper pots that are labeled "for decorative purposes only." Copper pots that are meant to be used as cookware are lined with stainless steel. Without this lining, dangerous amounts of copper can leach into your food.

Corning Ware:

- To clean: add four parts water to one part vinegar and bring to a gentle boil.

Enamel:

- To clean: Add salty water and let it sit for two hours. Cover the pot and bring it to a boil.

Stainless Steel:

- Rub stainless-steel pots and flatware with a vinegar-soaked soft cloth.
- If a stainless-steel pot is burned, add salt to warm water and toss in a few pennies. Shake repeatedly until the burned area lifts off.
- Remove any blue stain from the bottom by adding 1 tbsp (15 mL) tomato paste to hot water. Let sit and it will disappear.

Teflon or Nonstick Pans:

- Never use them over very high heat. Boil 2 tbsp (30 mL) baking soda along with ½ cup (125 mL) vinegar and 1 cup (250 mL) water for twelve minutes. Rub pan with olive oil.
- Always use plastic or wood utensils on these pans, as their coating will flake off with metal ones.
- Don't worry if your nonstick pans have a few scratches or nicks. Only dispose of them if there are significant chips, as this will affect their performance.

Pie Tin or Cookie Pans:

- Dip a piece of raw potato in cleaning powder (baking soda or any other nontoxic cleaning powder) and scrub clean.
- Rub the pan with vegetable oil and bake it in a moderate oven. When loosened, cool and wipe clean.

POULTRY

- Red bones in poultry means that it's been frozen at one time.
- Never stuff a bird ahead of time. Wait until the last minute.
- Olive oil, lemon and a rubbing of garlic make a crisp, tasty skin.
- Cook in the oven at 425°F (218°C) for thirty minutes, then reduce heat to 350°F (176°C). Add twenty to thirty minutes to total cooking time for a stuffed bird.
- Chicken: twelve to fifteen minutes a pound (.45 kg)
- Turkey: twenty minutes a pound (.45 kg.)
- Duck: twenty to twenty-four minutes a pound (.45 kg)
- Goose: twenty to twenty-five minutes a pound (.45 kg)
- Poke it in the thigh with a fork to test for doneness. The bird is done when the juice runs clear.

POWER FAILURE

- In a blackout, turn off all appliances except one lamp.
- Leave a transistor radio working to keep up with the news.
- Don't open the refrigerator unless necessary; food will stay frozen up to forty-eight hours, but don't refreeze food. Use it within a few days.
- Close off most rooms and confine your activities to one or two rooms.
- If the house will be without electricity for some time, drain the pipes to keep them from freezing in very cold temperatures.

Emergency Rations to Keep on Hand:

- bottled water
- canned ready-to-eat food
- dry snacks
- a store of candles and matches, and a flashlight. Know where these are so you're not fumbling around in the dark trying to find them.
- If you have advance warning of a storm, fill every available container with water for drinking. Turn the fridge to the coldest setting possible.

PRESERVING FOOD (See Canning Food; Drying Food; Freezing Food.)

PRICE TAGS: Removal

- Remove the sticky goo left over from price tags by applying baking soda with a damp cloth to the surface. Rub off.

(*See also* Decals.)

PRUNING

- Evergreens should be pruned in June. Never trim more than two-thirds of the new growth and don't go back to the old growth or you'll kill it off. Cut out dead interior branches.
- Prune spring-flowering shrubs after they've finished blooming in late spring. Cut back by about a third.
- Prune late summer-and fall-flowering shrubs in the spring.
- Or try to remember this: prune fall-flowering shrubs in spring; prune spring-flowering shrubs in fall.
- Cut out dead or diseased wood.
- Cut on a slant with a clean pair of secateurs or shears, facing away from the direction you want a new branch to grow.
- Remove lilac suckers at ground level.

RABBITS: As Pests

- If wild rabbits are attracted to your garden, place plastic fencing around tender trees and shrubs to keep them from nibbling away at these treats.
- Put up chicken-wire fences dug 6 in (15 cm) into the ground and to at least a height of 2 ft (.6 m).
- A double row of onions might discourage them.
- Protect young plants by placing plastic milk cartons with the tops cut off over them.
- Sprinkle hot cayenne pepper around tempting plants.
- Don't leave places such as brush piles for them to rest in.
- To trap a rabbit: place rabbit droppings near the opening of a trap. Release a rabbit into the wild many miles away.

RABBITS: As Pets

Rabbits can live for fifteen years, so be sure you really want one. Bunnies should be at least four weeks old before you take them home.

- Never pick up a rabbit by its delicate ears. The loose skin over the shoulders is best with the other hand under the animal.
- They molt during the fall. They may eat some of their droppings, but it won't hurt them.

Feeding Rabbits:

- Buy rabbit pellets from a pet store. Adults receive 5 oz (140 g) a day, young rabbits as much as they can eat.
- On occasion, give it carrots, grass, lettuce and almost any green except cabbage. Too many greens will cause diarrhea.

To Housebreak a Rabbit:

- Enclose in area where you want the rabbit to relieve itself. Put a kitty litter tray wherever it relieves itself the first time. This is where it will return. Clean litter daily.
- Protect furniture and make sure there are no electric cords available for gnawing.
- Build a hutch with a removable litter drawer.
- Clean out daily and scrub out once a week with disinfectant.

RACCOONS

In the Garden:

- Make a floppy wire fence: attach chicken wire to stakes and leave the top 12 in (30 cm) free. It will bend backward when the animal tries to climb over.
- Sprinkle black pepper or cayenne around vulnerable plants.
- Always remove raccoon droppings (similar to cat droppings that haven't been dug in) with a shovel. They may be fatal if they get on your hands.
- Grow enough food for the raccoons or put crushed corn around corn plants and perhaps they'll leave them alone.
- Plant a thick border of nasturtiums around your vegetable patch.
- Build a frame around your vegetable beds and put a screen over it.
- Fill net bags with your own hair (or anyone else's you can get hold of) and hang at intervals around your vegetable patch. This may discourage them since it demarcates your territory.

In Your House:

- To prevent them entering your home, set a radio on a timer that goes on and off in areas where they might enter the house—the attic for instance.
- If they nest in your attic, call a qualified exterminator who will trap the raccoons and release them in the wild.
- If you are not wary of trapping them yourself, set a trap baited with canned fish. Take them to a spot at least 20 miles (32 km) away or they'll be back before you know it. Be careful. Raccoons can carry rabies.

RADIATORS: Car

- The most environmentally friendly way to clean a radiator is to take it to a reputable service station where they will dispose of the chemicals properly.

- If you must clean a radiator yourself, drain it through the plug in the bottom. Try to contain spills. Flush it out with a garden hose and refill with antifreeze.

RADIATORS: House

Keep radiators working at top performance to get your best energy savings out of them.

- If you have a hot-water heating system, wrap the feed pipes (the ones that lead into the radiator itself) with insulation. This will keep them warm and not allow heat to escape inefficiently.
- Bleed the air out of the system before starting up for the year: open the valve with a screwdriver. With the heat on, hold a cup under the valve. Open it slowly. When air has finished escaping and very hot water spurts out, close valve quickly. This will eliminate chances of the pipes knocking when you turn on the heat.
- Increase heat radiation by putting a sheet of aluminum foil on the wall behind a radiator.
- To increase the humidity in your house and bring it up to a comfortable level, put pans of water on the radiators with a piece of charcoal in each container to keep the water sweet.

RATS

- Ratproof your house by blocking all holes in cement foundations and cracks.
- Pack steel wool into holes in plaster walls and replaster.
- Make sure there are no gaps around pipes and wiring holes.
- Use heavy screening around windows and any chutes or vents going from inside to out.

RECYCLING

In the ecological house you will recycle as much material as possible. Find out where your local recycling branch is located to get any rules they have. To be a good recycler, you are going to end up with piles of stuff on your property waiting to be picked up. This is a small penalty to pay for reducing your personal waste. Buy products that have minimal packaging. Somehow cheese in slices with bits of plastic between them has not seriously enhanced our way of life.

- Buy as many goods as possible made from recycled materials.

- Reuse plastic bags, aluminum foil plates, gift wrap, string and containers for as long as you can.
- Return clothes hangers to the dry cleaners.
- Donate used eyeglasses to service organizations.
- Trade in your old car battery when you buy a new one. Think about getting a rechargeable battery in future.
- Reuse bleach bottles and other large plastic bottles the following ways: half fill with water and use as bumpers on your boat; use as boat markers; cut out in the shape of a sand scoop, which can be used of myriad things, including a pooper scooper.
- Cut the top off a plastic bottle and use as a funnel.
- Cut up plastic bottles and use as plant identifiers in the traditional shapes; keep them for covering tender plants in the spring; punch holes in the bottom and use them to start seeds in.
- Use old mustard squeeze bottles as laundry sprinklers, or to hold liquid detergent.
- Reuse an old typewriter ribbon. Put it in a bag with some ink and store for a couple of weeks in a dark place.
- Have old tires retreaded or take your worn tires to a rubber-reclaiming plant if you can find one.
- Reuse the plastic tabs on bread for all sorts of scraping chores such as getting burns off pots.
- Always save gift wrap to use again.
- Wash and keep plastic foam trays (the kind used under meat from the supermarket). Use them to pack lunches in; back photos; put between plates for safe storage or moving.
- Patronize stores that will take back plastic bags and that do not stock nonreturnable bottles.
- If you are sending fragile goods, pack them in popcorn rather than Styrofoam chips.
- If your community hasn't started a newspaper-recycling program, put the pressure on politicians to get one going.

Recycling at the Office:

- Get your office or workplace to convert to using recycled paper (bond and computer). It is good-looking, though more expensive than regular paper. Get into a fine paper-recycling group.
- Don't use styrofoam cups—bring your own mug from home.
- Have a can-recycling bin near a pop machine.

(*See also* Home Offices.)

REFRIGERATORS

- Test the condition of the refrigerator gasket by closing a page of newspaper in the door. If the seal is tight, the page will rip when you try to remove it. If it slides out, clean the gasket or replace it.
- To keep the refrigerator fresh, always leave an open box of baking soda in both the freezer and cold sections. Change every month or at most two months. A piece of charcoal used for potting plants will also do the trick.
- Check for old leftovers weekly and put them in the compost or throw them out.
- Try a little vanilla extract to improve fridge odors.
- Wipe the refrigerator with vinegar or baking soda and water. This will prevent mildew and make it very clean.
- If you have an older machine that isn't self-defrosting, try this trick: after defrosting the freezer, rub down the inside with glycerine. Future sheets of ice will loosen much more easily.
- To improve efficiency, clean the coils on the back of the fridge every six months. Use a cloth with ammonia on it or a vacuum cleaner. Make sure the area is well ventilated, and unplug the fridge before you start cleaning the coils.
- Always make sure the fridge is at least 2 in (5 cm) away from the wall.
- Lowering the temperature on the fridge will save electricity and money: food compartment should be 38°F to 40°F (3°C to 4°C); the freezer should be 0°F to 5°F (−18°C to −15°C).

RIPENING FRUIT AND TOMATOES

- To hurry up the ripening process in fruits: put the green fruit in a plastic bag punched with holes.
- Direct sunlight makes green tomatoes mushy. Let them ripen with stems up in a place out of direct sunlight, or put in a brown paper bag and leave in a dark place. Check regularly.
- Harvest cherry tomatoes and late tomatoes with the stems still on the fruit and put in a dark place. They'll be tasty for weeks if you have enough of them.
- Wrap green bananas or green tomatoes in a wet tea towel and put them in a paper bag.

(*See also* Avocados; Tomatoes.)

ROOFS

- Wear sneakers on a roof.
- Don't work on days when there is a big wind, when it's too hot (you want to avoid sunstroke) or when the roof is wet.
- Be careful of the old roofer scam: they call on you and point out all sorts of drastic things that they can fix for "only X-hundred dollars." Always get a professional who comes highly recommended or one who is a member of a roofers' association. Get more than one estimate.
- Flat roofs: make sure water isn't accumulating in one spot.
- If asphalt shingles curl or if a shingle is torn, take it out and replace it.
- Check the flashing or water-resistant material that protects the joints. Fix any leaks with caulking. Even the tiniest hole should be covered with roofing cement.
- Clean rusty spots with a wire brush and cover with metal primer.
- The easiest way to remove asphalt shingles is with a flat 6 in (15 cm) shovel. You could also use a pitchfork—the kind with flat, heavy tines. Make sure the tines stay parallel with each other. If any should bend, straighten by putting it in a ¾ in (2 cm) water pipe and pulling back into shape.
- To protect shrubs and grass from the mess of stripping a roof, place an old sheet of plastic or cast-off shower curtain liner over them. Staple to a horizontal 2-by-4.
- To test how waterproof your own roof repairs are, pour a pot of water over the area. If bubbles form that means the roof still has holes. On a shingle roof, this won't work—you'll have to wait until the next rainfall.

RUG CLEANING (See Carpets.)

RUG REPAIR (See Carpets: Repairing.)

RUGS: Area

- If you have a tufted area carpet that slides around, apply a few coats of shellac to the back and let dry between each coat. This treatment is not recommended for woven or knotted carpets.
- Use a special underpad coated with rubber.
- Apply latex backing yourself with a stiff piece of cardboard or a spatula.
- Put a rubber bath mat under a slippery outdoor carpet in icy weather. It will definitely stick.

SALADS AND GREENS

- To keep oil and vinegar dressing from separating, add ½ tsp (2.5 mL) raw egg white and shake vigorously.
- Wash greens and put them along with a stainless-steel utensil in a plastic bag and keep in the crisper.
- To dry a big load of greens for a party, put them in a clean pillowcase and put the whole thing into the washing machine on the spin-dry cycle for a couple of minutes.

SALT

- Add ½ tsp (2.5 mL) raw rice in a shaker to keep it from becoming moist.
- If you've put to much salt in the stew, add a raw potato or apple and then remove it once the salt has been absorbed.

SALT DEPOSITS

- Wash your car regularly. (See Cars.)
- To get salt out of carpets, let the area dry and go over with a stiff wire brush.
- To remove salt stains from shoes and boots, wipe off any loose muck, then dab with a solution of equal parts of vinegar and water. Let them dry before you polish.
- Wash salt-damaged clothes by adding 6 tsp (30 mL) vinegar to 2½ cups (625 mL) water.

SAWING

- Let the saw do the work. Use a stroke that's comfortable for you.

- To keep plywood from splitting, put a strip of masking tape just where you want to start sawing.

SCRATCHES

- Remove scratches from a glass-topped table by rubbing them with toothpaste. Rinse and clean as usual.
- On appliances, try automobile touch-up paint.
- Or use white correction fluid and sand lightly between applications until it's even with the whole surface. Don't buy correction fluid particularly for the job. It's only recommended if you want to use up what you have.
- Mirrors: put a piece of foil over the scratch on the back and shellac over it.
- Keep your cat from scratching furniture by putting double-sided sticky tape over the designated spots. Or give it a spritz of water every time it comes back to the scratching spot.

(*See also* Furniture Repairs.)

SCREENS

- If a screen shows signs of rust, rub both sides with a stiff brush soaked in turpentine.
- If you happen to still have brush-type hair rollers somewhere around the house, reuse them. Rub over a screen to pick up lint or dust.
- Hose down screens outside, then scrub with phosphate-free detergent and water with a soft brush. Hose again. Make sure the window's closed!
- To make a patch, cut a piece of similar screening 1 in (2.5 cm) larger than the hole and glue it in place with a clear cement.
- Metal screens: unravel a few strands on each perimeter of the patch, turn up the ends and push them through the screen. Fold flat on the back.
- On plastic screens with only a small tear, use clear silicone adhesive and poke it into place with a pencil point.

SCREWS

- Disguise screws by sinking them slightly below the surface of the wood on an attractive piece of furniture or a bookcase. Patch over the depression above the screw head with a matching material—a wood plug or piece of doweling.

- Keep this rule of thumb in mind: tight to the right, loose to the left.
- A screwdriver often won't slip if the blade is coated with a bit of chalk.
- Tighten a screw by sticking a wooden match into a screw hole and replace the screw. You could also use a piece of solder.
- Screws that keep coming out can be anchored with a drop of shellac under the head before tightening.
- A screw cap that won't budge can probably be moved using a nutcracker.
- A screw that is too tight will sometimes come loose if you heat the edge of the screwdriver. Or put on a few drops of hydrogen peroxide and let it sink into the screw, then try again.

SEALING WINDOWS (See Windows.)

SECURITY: Apartments

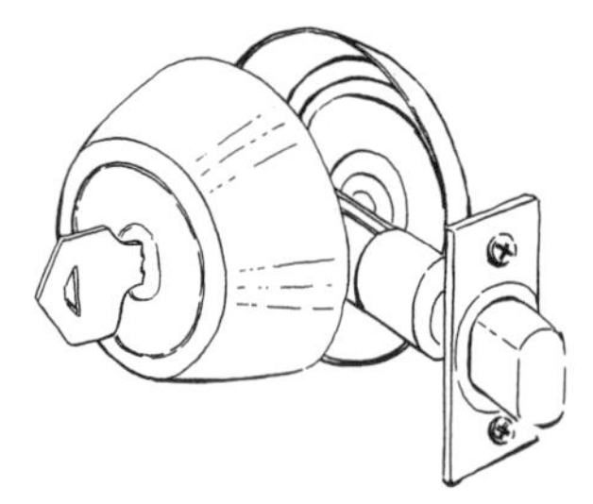

- Check first with the superintendent, then change the lock on the front door as soon as you move into an apartment. Install a good quality dead bolt lock and be sure to lock it when you leave the apartment.
- If there isn't a peephole, install one yourself.
- Repair any cracks or spaces in the door frame so that the door fits snugly.
- Be sure to lock windows and balcony doors. Don't assume that being on an upper-level floor means that intruders can't enter your apartment.
- Don't let others know when you are away for the weekend. For example, don't leave a note for the delivery person on your front door or in the lobby. Let a trusted neighbor know you're going to be away.
- If you have a telephone-answering machine, leave a deliberately vague message. It's advisable for women living alone to imply that there is more than one person living there, i.e., "We're busy at the moment, so please leave a message."

SECURITY: Houses

Most home insurance policies will become invalid if you leave your house completely unattended for more than a certain number of days. You should read the policy before you go away and make sure your house is checked on a regular basis by a friend or a neighbor.

Most burglaries take place during the day. Remember the following when you leave the house:

- Make sure that all doors and windows are shut and locked. Investing in deadbolt locks is wise.
- Leave lights on in selected parts of the house or have them hooked up to a timer that will go on in a logical sequence through the house (i.e., bedroom lights go on at a suitable bedtime).
- Lock up all ladders; make sure any attic or roof doors are also securely locked.
- Make sure that the skylights are secure and can't be reached by a drainpipe.
- Check the garage and make sure that all entries are securely locked.
- Talk to your local police department and find out the modus operandi of local thieves and the best ways to guard against them.
- Make sure all exterior doors are of solid construction—either wood or steel—with undamaged, close-fitting frames. Make sure double glass patio doors have special locking bolts to keep them closed securely.
- Place metal grilles over all basement windows. But be sure that they won't prevent escape in case of a fire.
- Don't grow heavy shrubbery close to the house. It's just right for hiding in.
- Be cooperative with neighbors. Exchange house-sitting duties such as collecting papers and trimming the lawn when on holiday.
- Etch identification numbers onto valuable items, so that if they are stolen, they can be easily claimed.
- Don't leave unlit areas around the house. Leave a porch light on during the night.

(*See also* Burglar Alarms.)

SETTING A TABLE

Setting a proper table follows a time-honored pattern. This is assuming there is soup, fish, then a meat and vegetable course, followed by salad and dessert.

- You work from the outside in to the plate.
- Forks always go on the left. A small fish fork used for seafood cocktail however, would be placed to the right of the plate.
- To the right of the plate from the outside in: the soupspoon, butter knife, fish knife and dinner knife (with the sharp edge facing toward the plate).
- To the left of the plate: the napkin, fish fork, dinner fork and salad fork. Of course if you serve salad before the main course, you would reverse this.
- The dessertspoon and fork can be placed at the top of the plate with the spoon pointing left and the fork to the right.
- Salad plate goes to the left of the forks, coffee cups to the right of the knives.
- Remove plates from the right, serve from the left.
- Water glass is closest to the plate, wineglasses outside this. Wine is always served from the right during the course, not between courses.

SEWING (See Darning; Mending Clothes; Patching Clothes.)

SHARPENING

Axes:

- File both surfaces with a medium-cut mill file held flat against the blade.
- Hone edge with an oiled, round ax stone.

Knives: ***(See Knives.)***

Scissors:

- They don't necessarily improve with sharpening. Try tightening the pivot screw first—this may do the trick.
- Move the blade diagonally across the whetstone. Wipe clean then open and close to remove the burr.
- Get the rust out by soaking for a few minutes in household ammonia and wipe clean. Be careful not to breathe fumes.
- To get a superficial sharpening, cut up a piece of sandpaper with dull scissors. Or cut up steel wool pads if you haven't any sandpaper.

Tools:

- Start by restoring the beveled edge: use coarse-grit stone and dip the blade in water frequently.

- Maintain the angle of the original bevel of the tool and slide the blade back and forth firmly.
- Finish with a fine-grit stone.
- When you have a shine and you can feel an edge of metal or burr, you've done it.

SHELLFISH

- Clams and oysters: wash and place in freezer for an hour—they'll open up readily.
- Never overcook any shellfish; they turn rubbery very quickly.

SHOCK

After an accident, look for the following symptoms of shock:

- anxiety and restlessness
- pale or blue-gray color in the skin (check lips, fingernails and earlobes)
- pale, sweaty or feeling unwell
- faintness and nausea
- delirium
- rapid pulse and breathing
- thirst—but do not give anything to drink.

Until medical help arrives do the following:

- Make the patient lie in a semi-prone position, feet raised slightly from 6 to 12 in (15 to 30 cm). Keep head to one side and loosen clothing. Keep warm but not hot (don't use hot-water bottles).
- If the person becomes unconscious, get help or rush to an emergency department.

SHOCK, ELECTRIC (See Electric Shock.)

SHOE RACKS

- Keep your shoes in order by mounting a piece of doweling or a curtain rod a few inches (centimeters) off the floor of the closet.
- Store shoes in liquor store boxes—the ones with the dividers.
- Utilize an abandoned towel rack by mounting it near the floor of your closet and storing shoes there.

SHOES

- Stretch tight shoes by balling a couple of sheets of wet newspaper into the toe of each shoe (after you've squeezed out most of the moisture).
- Stick a long baking potato in the toe of the shoe and leave it for a couple of days.
- Dry wet shoes by hooking heels on the rungs of a chair. This will provide good air circulation.
- Improve dried shoe polish (the kind in glass containers) by putting it in a cup of water and sticking it in the microwave for forty-five seconds.
- Always use a shoehorn and if you can't find one, substitute a soupspoon or the top of a plastic container with the edges cut off.
- If you're afraid of slipping in brand-new shoes, just go over the soles with a piece of sandpaper to rough them up.
- Dip broken shoelace tips into clear nail polish. Let dry and rethread.

Shining Shoes:

- Buff white shoes with a crumpled ball of wax paper after they've been polished.
- If you wear white shoes to the office, wrap desk and chair legs with white adhesive to keep them from getting scuffed.
- To get an even finish, rub with a raw potato before applying white polish.
- Milk will remove finger marks from patent leather—just apply and let dry, then shine.
- Put a well-worn old sock around your hand and use this to polish your shoes.
- After you've put polish on your shoes, buff with a few drops of lemon juice.
- Clean brown shoes with the inside of a banana peel.
- Spray canvas shoes with fabric protector to keep them looking new.
- Clean rope-trimmed shoes with a toothbrush dipped in rug shampoo.
- Spray tennis shoes with starch to keep them looking good-as-new.
- Get rid of those shiny spots on your blue suede shoes by touching up with a spot of vinegar. Let dry and brush.
- An emery board will get rid of rain spots on suede shoes.

- Brush up suede shoes with an old pair of panty hose.
- Revive shiny suede shoes with a dusting of talcum powder and gently rub in a circular motion with very fine steel wool or sandpaper. Remove the talc with masking tape. Brush.

(*See also* Boots; Salt Deposits; Stain Removal: Fabrics.)

SHOWER DRAINS (See Drains.)

SHUTTERS

- If a shutter sags, try tightening or adjusting screws or hinges. If this doesn't work, replace them. Or remove nails and screws and fill holes with plastic wood, matchsticks or a piece of wood dipped in glue before replacing.
- To protect wooden shutters and blinds from rot, warping and mildew, tack a narrow strip of metal across the top.
- To replace a shutter: each shutter should cover half the width and the full height of the sash opening. Mount hinges on the face of the window frame—the shutters should swing inward to the center.

SIDEWALK MAINTENANCE

- Check for and repair hairline cracks annually.
- Remove oil and grease by covering with portland cement or hydrated lime. Sweep away and reapply until stain disappears.
- Remove stains by mixing one part sodium citrate to six parts water, combined with six parts glycerine. Mix with hydrated lime. Form a paste and spread on stain.
- In winter, sprinkle biodegradable kitty litter instead of destructive salt all over your sidewalks. It will give an abrasive surface and won't destroy your walk. (*See also* Ice Removal.)

To Patch a Sidewalk:

- Widen crack to 1 in (2.5 cm) and roughen. Wear safety glasses when chipping out old concrete. Clean thoroughly, apply patching concrete and allow to dry for two hours before troweling.
- Protect all patches from sunlight for two days.

SIDING: Painting (See Painting: Outside.)

SIEVES

- To strain effectively, put a clean handkerchief in the bottom of the strainer. Always wash the strainer and handkerchief after using.
- If you put food in the processor for a second or two, it will go through a sieve much more easily.

SILK (See Hand-washing Clothes.)

SILVER CLEANING

- Put a piece of aluminum foil in a glass or enameled dish that contains 1 quart (1 L) boiling water to which 1 tbsp (15 mL) washing soda has been added. This causes a reaction between the silver and the aluminum and cleans in an instant. Don't use this on carved or raised designs with a "French" finish.
- Put your silverware in one of your recyclable aluminum pans. Fill with hot water and leave it overnight. A catalytic reaction will take place.
- Add 1 tbsp (15 mL) each of baking soda and salt to 1 quart (1 L) water. Put silverware in pot and let boil for three minutes. Polish with a soft cloth.
- Rub stubborn parts with white toothpaste and a toothbrush, then polish. Never reuse the toothbrush for your teeth, even if you boil it back to health. Some silver will penetrate the brush and eventually your teeth.
- Remove egg stains by rubbing with a dry cloth sprinkled with salt. Rinse and dry immediately.
- Make a polishing cloth this way: soak a piece of flannelette in milk for thirty minutes. Hang to dry without squeezing the liquid out.
- For sterling silver, use a typewriter eraser on the worst spots.
- For storing silver: put a piece of alum in the drawer or silver chest when you put it away.

SILVERFISH

- Sprinkle Epsom salts along the baseboards, in crevices and in corners, but be careful if you have kids or pets.
- Use an environmentally-friendly insecticidal spray or powder in the moist areas where silverfish lurk. They may live behind wallpaper or in bookshelves.
- Sprinkle borax around shelves.
- Mix equal quantities boric acid, sugar and plaster all over the affected areas.

- Put slices of lemon (new or old) in areas where they lurk.

(*See also* Insect Control.)

SINKS

- Sink stains: wet the area and sprinkle with baking soda. This will also leave it shiny.
- Dampen a cloth with vinegar and wipe the sink, or use club soda to make it shine.
- Try using a piece of ribbed wool with baking soda to shine up a stainless-steel sink.

(*See also* Kitchens: Cleaning and Maintenance.)

SKIN CARE

- Check labels to see if the product is biodegradable.
- Buy products with the least amount of packaging.
- Write to the manufacturers of overly packaged or non-recyclable packaged items.
- Choose soaps that have no artificial scent.
- Avoid disposable products, like razors.

SKYLIGHTS

- To inspect a skylight, look at caulking beads and roofing cement seams.
- To make your skylight vandal-proof, construct a screen over the outside. Have this done when the skylight is being installed; lay ½ in (12 mm) screening over the glass, then install an aluminum angle and caulk.
- To clean a skylight, the easiest thing to do is hose it off on the outside and use a homemade glass cleaner on the inside.
- To clean Plexiglas bubbles, don't use anything abrasive or you'll scratch the surface. Get a polymer protector, which will act like a wax without dulling the surface. Use the cleaning product recommended by the manufacturer of the skylight. Ask if a weak vinegar and water solution will do the trick.
- Small scratches can be covered up with car wax.

SLUGS

- The best thing you can do is partially bury a container filled with beer and leave it almost covered. Slugs from all over the area will come in, get drunk and drown.

- 1 tbsp (15 mL) baker's yeast to 3 oz (85 mL) water is just as effective if you've run out of beer.
- Grapefruit rinds placed hollow side down for three days at a time will attract the insects; remove.
- Put a barrier of diatomaceous earth around vulnerable plants.
- Ring precious plants with fireplace (not barbecue) ashes.

SOAP SAVING

- Make a bar of soap last longer by putting it in with your sheets and towels or lingerie. Smells nice, stays hard and will discourage moths.
- Put all your old bits of soap into a bag made from a wash-cloth. Close it with a drawstring or Velcro.
- Grate up all your leftover soap and you'll have your own soap flakes.
- Melt all the bits of leftover soap and pour into a mold such as a travel soap container (one without holes in the bottom). Let it dry out for several weeks.
- Put all your soap bits into a squeeze bottle and fill with water for a liquid hand-washing soap.
- Grate all those little leftover bits of soap together with some hot water—just enough so they'll adhere to each other—and press them into a mold. A drop of your own perfume will make it special. You'll have an almost new bar of soap.

SOCKS

- Restore white socks to pristine condition by boiling them in lemon juice and hot water.
- If you lose the mate to a sock, wash and store in the freezer for a instant cold compress for tired eyes.

SOUNDPROOFING

If you are having trouble with keeping down the noise level in your house, try some of the following:

- Use thick carpeting, heavy curtains and lots of upholstered furniture.
- On a common wall, try putting up frames containing gypsum-board paneling on each side. Leave ½ in (12 mm) airspace between the frame wall and the common wall to help deaden sound.
- Look for any cracks through which sound can leak—make sure ceiling, floor and walls are well caulked or sealed. Fill

gaps around doors, since most sound comes around a door, not through it.

SPICES

- To bump up the pungency of curry powder, heat it in the oven for a few minutes before adding to the dish.
- Add cinnamon sticks and vanilla beans to a jar of sugar to make a tasty addition to your morning coffee.
- Ginger: wrap it in plastic and keep it in the freezer, ready to be grated into dishes: 1 tbsp (15 mL) equals ½ tsp (2.5 mL) powdered ginger.

SPLINTERS: Removing (See First Aid.)

SQUEAKS

- Furniture drawers: empty, turn upside down and apply a greaseless jelly for the purpose. Or rub the stub of a candle or a cake of paraffin where wood moves against wood.
- Windows: rub paraffin wax or a cake of soap along the tracks in which the sash moves.
- Aluminum awning window: you'll need a special lubricant for this job—it needs to be sprayed on all moving parts. Take it apart once a year and apply with grease.
- Door locks: powdered graphite will help, but don't use oil. Blow the powder into the lock through the keyhole.
- Air conditioner: a squeak means the blower motors need lubricating—check your manual before you call for servicing.
- Squeaky floors can be fixed if you pour talcum powder between the boards, or run glue into the cracks.
- Squeaky stairs: brush the underneath with baby powder, wipe the excess with a damp cloth.
- Squeaking hinges can be silenced if you run a lead pencil over them. If this doesn't work, rub with a bar of soap.

SQUIRRELS

If squirrels become a nuisance in your garden, try the following:

- Provide a little extra food for them in one part of the garden. They can be quite lazy.
- Scatter blood meal around plants, especially bulbs, which they love. It's a good fertilizer, but has to be replaced after each rain or watering.

- Sprinkle cayenne or black pepper near anything they like to eat. Pepper is far more effective than using mothballs, which are toxic anyway.
- If your cat wanders outdoors, it may act as a deterrent. Dogs are often more effective because they like to mark a territory with their scent, and squirrels are territorial.
- Hang hanks of your hair in bags around the garden to keep squirrels at bay.

STAIN REMOVAL: Fabrics

When mixing your own ecologically-sound stain removers, be sure to follow these rules of thumb:

- Never mix ammonia and chlorine bleach or you'll produce deadly fumes.
- Never mix chlorine bleach with vinegar or you'll produce deadly chlorine gas.
- Once you've stained something, test to see if colors are affected. Select a small, inconspicuous area and apply a bit of the removal solution.

Animal Stains on Carpets:

- Scrub the stained area with a soft brush soaked in a mixture of laundry soap for delicate fabrics and cool water.
- Sponge with a solution of 1 tbsp (15 mL) borax in 1 pint (500 mL) water.
- Spritz the area with soda water. (*See also* Animal Odors.)

Ballpoint Pen: See Ink in this section

Beer:

- Rinse in cool water and apply a few drops of lemon juice or vinegar in 1 cup (250 mL) water, then wash regularly.
- If stain persists, put washables in warm water with a little ammonia.

Beets:

- For anything that's washable, soak a piece of bread in cold water and put it over the stained area. Once the bread has absorbed the stain, wash as usual.

Blood:

- Gently sponge the stain with cold water or baking soda. Then sponge with a solution of salt and water. Blot dry with a towel. Repeat until the stain is gone.
- For white cottons and linens with stubborn stains, soak in cold water, rub phosphate-free detergent into the stain and rinse. Wash as usual.
- On unwashables: mix cornstarch with water until you have a thick paste and spread on stain. Brush off when dry.
- On carpeting: sponge the stain with phosphate-free liquid detergent, followed by clean water. Hydrogen peroxide will also help, but don't let it soak into the carpet—sponge off immediately.
- Or mix meat tenderizer and water to make a paste. Apply and let rest thirty minutes, then sponge with clear water.

Chewing Gum:

- Put item in the freezer to harden, then scrape. (*See also* Chewing Gum Removal.)

Chocolate:

- Rub with a borax and warm water solution before laundering: 4 tbsp (60 mL) borax to 2½ cups (625 mL) water.
- Or soak in cold water, rub phosphate-free detergent into stain and rinse.

Coffee, Tea:

- Immediately pour boiling water from 1 yd (1 m) in height over a fresh stain until it disappears, unless you are worried about delicate fabrics.
- Soak stain with borax and water, then wash as usual.
- Apply glycerine on the spot, leave on for fifteen minutes and wash.
- Whites: pour boiling water through stain. Bleach if necessary.
- Other washables: soak in cold water and work phosphate-free detergent into stain. Rinse and bleach if necessary.
- Coffee with cream: rinse or soak in cold water and work detergent into stain. Rinse and dry. Bleach if stain remains.
- On an old stain, use 1 tbsp (15 mL) borax and 20 oz (570 mL) water solution for fifteen minutes then wash.
- Spread an egg yolk mixed with a little glycerine over tea-stained wool. Leave for half an hour and wash in warm water.

Cosmetics:

- Ball up white bread and erase the stain from the fabric.
- Whites and other washables: sponge with cool water, work phosphate-free detergent into the stain, rinse with tepid water. Follow with equal parts of 6 percent hydrogen peroxide and water if there's a yellow stain.
- Make a paste of granular detergent and rub into the stained area, then wash.

(*See also* Lipstick in this section.)

Crayon:

- Put fabric between pieces of brown paper and press with an iron.
- Or do the same with paper towels.
- Rub the mark with toothpaste, leave for fifteen minutes and wash off.

Cream or Milk:

- Rinse in cold water, work phosphate-free detergent into stain and launder.

Egg:

- Cover with salt and let sit for an hour before washing.

Fruit Stains:

- Stretch stained area over a bowl and pour boiling water from a few feet above.
- Dab with eucalyptus oil and leave overnight. Wash with detergent and warm water.
- Wash delicate things with cool water, then dab with glycerine and leave on for one hour. Follow up with detergent and warm water.

Grass:

- Put biodegradable liquid dishwashing detergent on stains, let sit for ten minutes and wash.
- Or dampen the stain and cover with sugar. Leave it rolled up for an hour, then wash.
- If the stain persists, make a mild solution of chlorine bleach or hydrogen peroxide. Sponge with water, then apply a few drops of the solution. Repeat and let dry.

- On colored clothes, sponge the stain with a solution of glycerine and paraffin. Leave for at least an hour.
- This sounds weird, but try blackstrap molasses on the stain, then soak and wash.

Grease:

- Pour some baby powder on the spot, give it time to absorb the grease, then brush off.
- On suede, dip sponge in vinegar or club soda and brush with a suede brush in the direction of the nap.
- Rub fine oatmeal gently into the suede, brush off.
- Ball up a piece of white bread and use it like an eraser. (*See also* Leather.)

Ice Cream and Baby Formula:

- Rinse in cold water, then add unseasoned meat tenderizer to the area before you wash.

Ink:

- Try soaking with milk and rubbing the stain away.
- Or soak the fabric in milk over night and wash as usual.
- You can also try putting cream of tartar on the stain, followed by a few drops of lemon juice. Rub into the stain for a minute, then brush the powder away with a clean brush and sponge with warm water.
- Stain still there? Dab it with rubbing alcohol before washing.
- Alas, india ink is just about unremovable, but you can try rubbing the spot with kerosene first. This is a highly flammable product, so be careful with it.

Lipstick:

- Try rubbing gently with white vinegar. If that doesn't work, try eucalyptus oil on a damp cloth.
- Dab with glycerine. Leave for half an hour then sprinkle with soap powder and launder.

(*See also* Cosmetics in this section.)

Mildew Stains:

- Make a mild bleach solution of hydrogen peroxide and water and apply to the stain. (*See* Mildew.)

Milk and Milk Products:

- Soak in warm water immediately. If the stain has worked its way into the fabric, sprinkle on some borax and soap powder. Pour hot water over the stain and rub. Leave for ten minutes and wash.

Mud Stains:

- You can take out residual stains by rubbing a potato over them and then soaking the item before washing.

Mustard:

- Scrape off any excess. Apply glycerine and let stand for thirty minutes. Work in biodegradable liquid soap (not detergent) and wash in hot water.
- For delicate fabric, rinse with a solution of equal parts of alcohol and water. Apply hydrogen peroxide and rinse.
- Put a bit of white toothpaste on the spot and rub gently.

Oil:

- Rub glycerine into the stain and let it soak for an hour. Sponge clean with detergent and warm water.
- Engine oil can be removed by sprinkling talcum powder on it. Let stand for thirty minutes and brush.

Paint:

- Soak item first, rub salt and hot water on the stain and wash.

Perspiration:

- Get rid of stained area with phosphate-free detergent and warm water, then sponge with equal parts of ammonia and water.
- On colorfast fabric, use warm vinegar before you launder.
- Dissolve two plain aspirin tablets ground into powder in water and sponge the offending area to rid of any odor.
- Sponge with lemon juice or slightly diluted vinegar.
- On synthetics, make a mix of 1 quart (1 L) cold water and 4 tbsp (60 mL) baking soda.

Ring-around-the-collar:

- Dry shampoo will take out the worst of the stain.
- Wet the collar and leave in a saucer full of phosphate-free liquid detergent for fifteen minutes before washing.
- Put chalk over the ring and leave it for twelve hours, then wash.

Rust:

- Cover stain with cream of tartar, dip whole spot in hot water for about five minutes and launder.
- Rub lemon juice and salt on spot and leave it in the sun.

Scorch Stains:

- Place stained fabric in ice-cold water for twenty-four hours before you wash.
- On unwashables, use a solution of 1 tsp (5 mL) borax to 1¼ cups (300 mL) hot water and pour over spot. Sponge with clear water.
- On white fabric, sponge area with a solution of one part hydrogen peroxide to four parts water and put garment in the sun.
- Use the same solution and cover with a dry cloth. Press with a medium-hot iron. Don't let the iron touch the hydrogen peroxide or you'll get rust stains.

Shoe Polish:

- Mix one part rubbing alcohol with two parts water and sponge on colored clothes; use straight rubbing alcohol on white.
- Gently sponge the stain with straight phosphate-free detergent, then rinse. If the stain remains, dab with methylated spirits and follow with a detergent washing.

Tablecloth:

- To get rid of fruit stains like ketchup and jam, stretch the stained area over a bowl and pour boiling water through the material until the stain disappears. If it's particularly stubborn, you might need a follow-up washing with bleach.
- Get the yellow out of an old tablecloth by putting 6 tsp (30 mL) hydrogen peroxide in 1 cup (250 mL) water and sponge the stain, then wash.
- A yellowed linen tablecloth will come up white if you boil it in water with a cheesecloth bag filled with crushed eggshells, then wash.

Tar:

- Dab glycerine on wool and let it soak for an hour. Wash with cold water detergent and rinse.
- Test the fabric first, rub the spot with kerosene, then wash. This is highly flammable, so be careful.

Urine:

- Soak in cool water, rinse and work in detergent. If color changes, sponge with ammonia. If stain remains, sponge with vinegar.
- With more delicate fabrics, dilute ammonia with an equal part of water.

(*See also* Animal Odors.)

Vomit:

- Sponge with cold water, work in phosphate-free detergent and rinse. Or sponge with vinegar, then wash.
- Let cat vomit dry and just brush off.

Water:

- On fabric: soak fabric in a solution of four parts water to one part salt. Wash the stains thoroughly and rinse with fresh water.
- On silk or velvet: steam for a few seconds and pat dry.

Wax:

- Whites: put stained item in the freezer and scrape off the wax. Put on blotters of facial tissues or paper towel and iron.
- On carpets: put a piece of brown paper bag over stain and hat with a warm iron just above the pile, so that the paper absorbs the wax and doesn't drive it into the carpet. Be careful not to melt nylon carpets.

Wine:

- Red wine: pour baking powder over the surface, allow to dry and vacuum.
- If it's on a cotton tablecloth, pour salt on it immediately. Let dry and brush away or vacuum.
- On an old stain: use the mixture of 1 tbsp (15 mL) borax and 20 oz (570 mL) water. Soak for fifteen minutes and wash.
- On wool or silk: wet stain with cold water, then follow up with glycerine. Leave for one hour, sponge off with lemon juice and rinse.

(*See also* Bathrooms: Cleaning and Maintenance; Caulking; Furniture Repairs; Metal Stains; Plumbing Fixture Stains; Sinks; Toilets; Walls.)

STAIRS (See Banisters; Squeaks.)

STATIC CLING

- Overdrying clothes automatically leads to static cling. Remove from dryer when they are barely dry.
- To eliminate static cling from clothing, stroke the garment with a wire hanger. Do the same if your hair is affected.
- A small amount of hand lotion on panty hose will prevent static cling.
- Add vinegar to the final rinse water to reduce static.
- Starch slips to keep dresses from clinging in embarrassing places.
- Spray water on the inside surface of the hem if static cling develops while you're wearing an outfit.

STORAGE

China:

- Put cloth napkins or pieces of newspaper between each plate to prevent scratching.

Jewelry:

- To keep costume jewelry from rusting, put a piece of chalk in the box.
- Use egg cartons to store precious things in.

(*See also* China; Clothes Storage; Food Storage.)

STOVES AND OVENS

- Leave the oven door open after you use it to heat a room.
- Preheating is only required for baking. If you need to preheat, limit time to ten minutes.
- Use a lid when cooking on the stove, as it cooks food two to three times faster.
- Don't unnecessarily open the oven door to peek. Every time you do, the temperature drops about 25°F (14°C).

Cleaning and Maintenance:

- If your stove has a charcoal filter to absorb odors, remove it and heat in a 450°F (232°C) oven for thirty minutes. Clean the grease and dust filter with soap and water.

(*See also* Appliances: Cleaning.)

Oven Cleaning:

- To prevent spills, use a pan or roaster that fits the size of what you're cooking. Use a drip pan or cookie sheet to catch potential spills.
- If large spills occur, throw a layer of salt on the spot at once. When the oven has cooled down, wipe it clean. If this doesn't work, pour baking soda on the difficult spots and leave for fifteen minutes. Wipe with a damp cloth.
- A paste of baking soda and water will remove grease spills.
- To clean the entire oven, make a paste of 1 cup (250 mL) borax and 1 cup (250 mL) vinegar with a little phosphate-free detergent. Heat the oven at 400°F (204°C) for five minutes and turn off. Spread the paste all over the oven and leave on for at least an hour. Scrape the gook off with a plastic pot scraper and wooden spatula. It won't be perfect, but it will be safer than commercial sprays.
- If the oven is disgusting, warm it for twenty minutes, then turn off. In a shallow dish, pour ¼ cup (60 mL) household ammonia and fill it with water. Leave in the middle of the oven overnight. Be sure that there is plenty of ventilation since ammonia vapors can be toxic and can cause copper to corrode.
- Oven window: take a wet sponge filled with baking soda and rub well. Rinse off with clear water. You may have to repeat this since grease bakes on the glass quite effectively.
- Self-cleaning ovens: if there are particularly heavy, baked-on spills, use washing soda or dry dishwasher soap on a dampened sponge and wipe clean. Do not use abrasives, steel wool or commercial oven cleaners. Use the self-cleaning feature right after cooking so the oven is already hot.
- Broiler pan: heat pan and sprinkle with phosphate-free detergent. Cover with a damp cloth and let sit. Scour with a paste of baking soda and water.

STRIPPING FURNITURE AND WOODWORK

Paint stripping should be done wearing rubber gloves and in a well-ventilated room. Paint stripper is incredibly toxic. You'd be better off renting a heat gun instead (see Paint Removal). There are some new products on the market that claim to be nontoxic and environment friendly. They won't burn your hands and you can clean up after with soap and water. Be aware of the dangers of lead paint. Wear a good quality respirator and do a thorough clean-up when you re finished. If you can't because the furniture is intricate, try the following:

- Apply paint remover one dab at a time in a 8 in (20 cm) strip in one direction only. Add more every five or ten minutes for an hour. Leave each section on for at least an hour. Most people start scraping too soon.
- After the paint has bubbled, leave it a bit longer, then lift the paint (don't scrape) with a putty knife. Repeat until you hit the wooden surface.
- Once all the paint is off, clean off all bits with mineral spirits or varsol. Sand if necessary and finish (see Furniture Refinishing).
- Put all the material in a safe container and make sure it's put in the hazardous-waste disposal in your area.

SUEDE

- Water spots on suede: let dry, then rub with a brush, pencil eraser, fine sandpaper or even an emery board.
- Brush grease stains with oatmeal. Rub with a clean cloth until the spot is clean.
- Or knead some white bread until it's soft and rub it on the suede like an eraser.
- Remove any surface dirt, brush with a wire suede brush or dry sponge.
- Synthetic suedes: machine wash unless the lining has to be dry-cleaned.

(*See also* Stain Removal: Fabrics.)

SUNSHADES

(See Awnings and Sunshades.)

TAPES (See Cassette Tapes.)

(See Cassette Tapes.)

THAWING FOOD

- Ideally, frozen food should be put in the refrigerator and thawed overnight. It's the safest method and you'll be more likely to retain the natural juices. All meat and poultry should really be done this way.
- Allow eight hours per pound (.45 kg) for meat, three hours per pound (.45 kg) for poultry.
- Partially-thawed meat can be cut into fine slivers more easily.
- Thawing at room temperature: allow two to four hours for fruit, three to four hours for bread, two to four hours for sandwiches and cake.

For a Quick Thaw:

- Immerse fish, meat or poultry in cold water while still wrapped. This cuts thawing time to one-eighth.
- To oven-thaw: place turkey in a 325°F (163°C) oven for three hours. Remove giblets, stuff, then return to oven immediately to roast. Any size turkey will do.
- Bread and rolls: to keep crisp, place in a brown paper bag in a 325°F (163°C) oven for five minutes.
- Microwave thawing: for poultry, allow eight to ten minutes per pound (.45 kg) and turn twice.
- Thaw frozen food in original containers.
- Thaw fish in milk to get rid of any chemical odors from the water.
- Unsalted butter freezes well. Thaw for three hours in refrigerator.

- To thaw cheese (the hard kind does best), allow three hours in the refrigerator.
- Fried food should be thawed at room temperature and baked in a 400°F (204°C) oven.
- Broccoli tastes better if you partially thaw it, but all other vegetables should be cooked frozen. Reduce the time you cook frozen vegetables from one-third to half the cooking times.

(*See also* Freezing Food.)

TICKS

- If one lights on your skin, hold a flame close to its body and remove it when it pulls its head out of your skin. Or use a red-hot needle or alcohol to do the same thing.
- If your pets are infested, wash and give them a herbal rinse of ½ cup (125 mL) fresh or dried rosemary added to 1 quart (1 L) boiling water. Let it steep twenty minutes, strain and cool. Sponge the animal with this.
- Treat affected areas in your house with a safe insecticide.

(*See also* Insect Control.)

TILES

- When applying wax to vinyl tiles, use a very, very small amount of paste wax on a cleaning cloth and work in small areas—nothing that will take more than ten minutes to dry. Using a cloth, apply liquid wax in circles.
- You can remove scum from tile walls by mixing ½ cup (125 mL) borax to ½ gallon (2 L) hot water. Scrub with a brush.
- To attach a soap dish to tiled walls, use a silicone adhesive caulk or construction adhesive.

(*See also* Bathrooms: Cleaning and Maintenance; Ceramic Tiles.)

TOILETS

Toilet-bowl Stains:

- Start by sprinkling with baking soda. Form a paste and let it sit before scrubbing well.
- If the baking soda isn't enough, mix two parts borax to 1 part baking soda or washing soda and try the same treatment.
- If you have denture tablets, drop one in the toilet and let it foam for five minutes, then brush and flush.

- Leftover soda pop poured into the toilet bowl and left overnight will provide the same effect.
- A seriously brown toilet bowl can be completely restored to its pristine glory by plugging the drain with a bag of soil and putting in a few rhubarb stems and leaves covered in hot water. Let it stand overnight, then clean as usual.
- Get hard-water deposits out by leaving 2 cups (500 mL) vinegar in the bowl for an hour before flushing.

Toilet Drain: Clogged (See Drains.)

Toilet Flushing:
- A failed flush might mean a blocked drain. Give it a good workout with the plunger.
- Raise the float arm by adjusting the screw at the pivot end or bend the float arm.
- If the flap closes too quickly, shorten the chain by removing a couple of links.

Toilet Leaks:
- Unscrew the hinge pin. Lift out the valve plunger and take to a hardware store to get the right replacement ball-cock repair kit. It will have everything, including new washers.

Toilet Tank:
- One way to save precious water is to put an ordinary brick or a plastic container filled with water in the toilet tank.
- Consider investing in a low-flush toilet that uses only 1 quart (1 L) of water compared with 4 gallons (18 L) water in the ordinary toilet.
- If your toilet tank develops a crack, get it replaced. Attempting to fix it could lead to serious damage—it's razor-sharp if it splits open, and you'd very quickly have water everywhere.

(*See also* Bathrooms: Water Conservation; Condensation.)

TOMATOES
- Grow small tomatoes slightly larger than cherry tomatoes. Leave them on the vine and store in a cool cellar. They will last most of the winter.

- To ripen late tomatoes, hang tomatoes and vine upside down in the basement. Put something under to cushion the ripened fruit's fall.
- Tomato equivalents: 1 cup (250 mL) canned equals 1½ cup (375 mL) fresh chopped and simmered for ten minutes.

(*See also* Canning Food; Container Planting; Drying Food; Freezing Food; Ripening Fruit and Tomatoes.)

TOOLS

Basic Household Tool Kit:

- chisel: wood chisel
- clamps: pairs of quick-action adjustable clamps
- drill: ⅜ inch (1 cm) reversible, variable-speed set of high-speed twist bits
- hammer: 16 oz (.45 kg) claw hammer
- knives: utility knife and craft knife
- pliers: 8-in (20-cm) long locking-grip pliers; slip-joint pliers
- saw: 8 point crosscut wood saw
- screwdriver: five different heads in the handle
- wrench: adjustable wrench
- miscellaneous: 10-ft (3-m) long flexible-steel tape measure, combination square, framing square and a level

Storing Tools:

Look after precious tools, especially against rust. Put them in a warm, dry room, but if you must keep them in a garage or tool-shed do the following:

- Clean each tool and coat the metal parts of each one with a light machine oil. Apply a few drops of oil for a fine coat with a cheesecloth or old bed sheet.
- If rust does occur, remove with steel wool and a few drops of oil.
- Store tools so they won't get dull, especially chisels and files. Drive nails into a board spaced so that the tools can hang separately, or store each tool separately in a plastic bag.
- Store razor blades by wrapping them in masking tape if you misplace the original container.

(*See also* Gardens.)

TOYS

Buy safe toys appropriate for the age of your child.

- Never give a small child a jigsaw puzzle with small pieces.
- Stuffed animals should be checked for easily removable parts.
- Any toy with strings or cords, such as a toy telephone, is unsafe for a baby.
- Toys with small parts should not be given to children under three (squeeze toys with squeakers, stuffed animals with removable eyes and noses).

The following toys should be safeguarded against:

- small rattles for babies
- balloons for small children
- anything with points or sharp edges
- toys that break showing sharp edges
- electric toys for children under eight years old

(*See also* Childproofing Your Home; Play Safety.)

TRANSPORTING CHILDREN

By Bicycle:

- Never transport a baby under nine months by bike.
- Make sure the carrier protects feet and hands from the spokes.
- The seat should have a lap-and-crotch belt.
- The child should wear a protective helmet, elbow and knee pads all the time. Make sure the helmet isn't too heavy.
- Never use the kickstand when the child is in the seat.
- Practice riding your bike with a weight equivalent to the weight of your child before you start carting the child around.

By Car:

- If you buckle up, so will your children.
- All children will be safest in the backseat.
- Never hold a baby or small child in your lap.
- For babies and children under 40 lbs (18 kg), a car seat gives the most protection: for 18 to 20 lbs (8 to 9 kg) make sure the seat faces the rear; for 20 to 50 lbs (9 to 22.5 kg): car seat strapped to face forward; for 50 to 60 lbs (22.5 to 27 kg): booster seat with a harness. The car's seat belt is effective for children over 60 lbs (27 kg).

- If your car has been parked in the sun, check seat and belts to make sure they aren't hot before buckling in.
- Never leave a child unattended in the car, and always lock doors.

TRANSPORTING PETS

- Put a damp towel over a pet's travel cage. This will keep it cool and cut down static created by the movement of the car.
- Or attach a clothes hanger to the rear bumper to ground the car. This will also cut down on static, thereby reducing irritation for the pet.
- Get the pet used to the travel container by putting in a favorite blanket a few days in advance and let it sleep there.
- Absolutely never leave a pet in a car with all the windows closed. Leave it in the shade with the window cracked open if you must have an animal in the car.
- Don't feed the animal for six hours before you set out, and make sure it's had a bowel movement.
- Take water from home so it won't have to adjust to strange microbes.

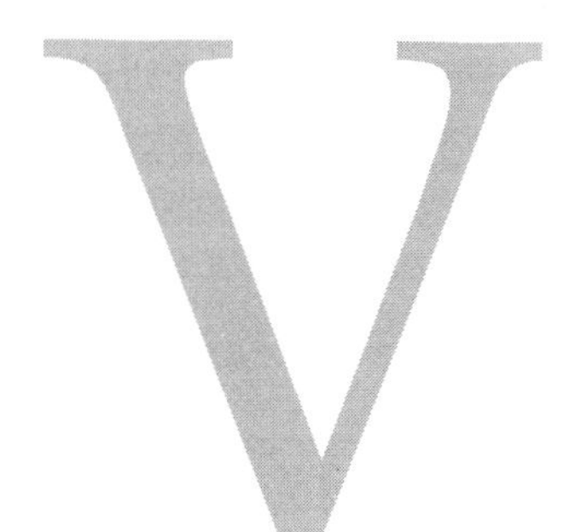

VEGETABLES

- To double a recipe for vegetables, only increase the liquids, spices and herbs by half the amount called for in the recipe.
- Lemon juice is filled with vitamins, so add a little to vegetables to keep the vitamin content as high as possible. It gives a more piquant taste as well.
- Boil water first, then add vegetables to the water. Don't over-cook.
- Always save vegetable cooking water (the exception to this is the water used to cook peas). Freeze and use at your convenience. You will have lost much of the vitamin content, but you will still have the flavor. Use as the base for stock in soups, gravies and sauces.
- You can steam most vegetables, and this is the best way to cook them.

(*See also* Food Storage; Pesticides: On Produce.)

VELVET

- Revive velvet by steaming it over a kettle and brushing it with another piece of velvet.
- To restore the nap, hold fabric over steaming kettle, add a little ammonia, then brush. Iron on the wrong side.
- Velvet made with cotton, acetate and viscose will have to be dry cleaned.

(*See also* Clothes Washing; Stain Removal: Fabrics.)

VENETIAN BLINDS

- Keep them clean by donning a pair of cotton work gloves or socks and rubbing each slat. You could dampen the gloves or socks in warm water and vinegar, as well.
- It's best to clean metal blinds in warmer months by taking them outside. Hose down and scrub each slat with biodegradable soap and water. Rinse with the hose and hang them on the clothesline, fully extended with slats horizontal, to dry. Try to avoid wetting the steel parts of the head rail.
- In colder months, immerse blinds in the bathtub to clean and hang up fully extended to dry. This prevents the connecting tape from twisting; however, it will probably shrink a little.
- Brighten up any blind by rubbing a damp sponge with a little white or colored shoe polish over the surface. Use the colored polish with caution—it could do damage.

VINYL (See Floors.)

WALLPAPER

- To calculate the number of rolls you'll need, measure the room (distance around the room times the height) and divide by 22—about the width of a roll. Take off two rolls for standard door and window openings.
- For patterns, start in the center of the room and work to the right—especially if you have to deal with a fireplace or windows.
- Before you start, cover grease spots with shellac and let dry.
- If you've never papered a wall before, you might want to use prepasted wallpaper. Start with the longest, uninterrupted stretch of wall.
- To cut a straight edge use a 6 in (15 cm) putty knife with a sharp blade, a utility knife or an X-acto knife.
- After papering a moist room, such as the laundry room or bathroom, paint all the joints with clear varnish.
- If you develop a bulge, slit it with a razor blade, insert a knife under the paper to release the air, then smooth with a wet sponge.
- Don't use metallic wallpaper around switches and sockets—they can conduct electricity.
- When papering a ceiling, make a platform of a long board supported by two stepladders. It's best to have a friend helping you. You might want to rent scaffolding. Take great care in assembling and using the scaffolding or the board-and-ladder arrangement.
- If the paper starts falling off, apply wallpaper paste to one corner of a piece of cardboard (or anything stiff). Blot excess then slip this under the wallpaper. Press to the cardboard, quickly peel wallpaper away from the paste and push back against the wall.
- Leftover wallpaper can be used to line drawers.

Cleaning Wallpaper:
Because wallpapers can differ, follow the manufacturer's instructions for cleaning and caring for your wallpaper.

Removing Wallpaper:
- Mix a solution of equal parts of vinegar and hot water and sponge or roller the paper thoroughly. Do this twice and peel off the paper.

Repairing Wallpaper:
- Tape a piece of new wallpaper over the damaged area. Match patterns carefully. Cut an irregularly-shaped patch from the new piece as you cut through the wallpaper on the wall. Remove the damaged piece and paste patch in place.

WALLS

To Clean Walls:
- Put a pair of panty hose over a mop head, or a sock at the end of a mop handle to get cobwebs off walls.
- Rub a stain with borax.
- Take folded recycled paper towels or blotting paper and press with a warm iron. Use fresh area each time you apply the iron.

To Maintain Walls:
- For a temporary repair, disguise small nail holes with a dab of white toothpaste. Smooth off with a damp sponge.
- Brush loose material out of crack and fill with Spackle. Wet the edges to help hold Spackle in place.
- Cracks: mix up white glue and baking soda into a paste.
- To replaster: if you add dry plaster to water, you'll get a lump-free mixture.
- Slow down plaster hardening with a little vinegar in the mix.

(*See also* Kitchens: Cleaning and Maintenance.)

WASPS

Wasps can be beneficial since they eat many other pests early in the year. However, they can also be a pain in the neck if you're eating outside.

- Set a place for them away from where you're working or eating. Make sure it's colorful and the wasps will congregate there.
- If you want to get rid of them, leave out a dish with highly-sugared grapefruit. They will sip until they drown.
- Keep wasps away from a hummingbird feeder with a cotton ball dipped in vinegar.

(*See also* Bees.)

WATERBEDS

- If you have a waterbed, make sure the bedspread is always on to keep the heat in. An unmade bed uses more energy to maintain the mattress's set temperature.

WATER: Bottled (See Bottled Water.)

WATER CONSERVATION

Hot water costs you about 20 percent of your total electricity bill in winter. It's like anything else in the ecologically-correct house—you want to save as much as you can on it. It is also important to conserve unheated water, particularly in the summer when droughts can occur.

- To conserve heat in your hot-water tank, lower the temperature from 150° to 130°F (65.5° to 55°C). Check to make sure that this will allow your dishwasher to run efficiently. You may find that you run out of hot water faster, though.
- Make sure that your water heater is kept clean by vacuuming around it regularly.
- To save energy, wrap the heater in a fiberglass wrap. This will hold in the heat and means more hot water in a shorter time. If you cannot afford a wrap, try taping batts together and wrap the heater in them. However, don't enclose the opening at the bottom of a gas water heater. Gas hot water heaters have a special wrap that should be applied by a professional.
- Turn off hot water when you're away, but be sure the house is heated—you don't want the water to freeze.
- When you run a bath don't run the cold water excessively.
- After a bath, leave the hot water in the tub until it cools to add moisture and warmth to cool winter air.

- Take a shower instead of a bath. Invest in a low-flow shower head for even more savings—up to 6,000 gallons (22,000 L) of water a year and 30 percent of the hot water you use for showering. You can put a pail in the shower with you and recycle the water—in the toilet, to water plants or for washing the bathroom.
- Do your laundry with cold or warm water instead of hot.

(*See also* Bathroom: Water Conservation; Water Heaters.)

WATER HEATERS

- If clogged up it will make a noise and might mean that sediment has built up. Open the drain valve at the bottom of the tank and let it run out until the water is clear. It's absolutely essential to turn off the electrical supply to the heater when you do this.

(*See also* Water Conservation.)

WATER MARKS (See Furniture Repairs.)

WAX: Candle

- Candlewax seems to drop wherever it's used. When it overflows on to the candlestick, put it in the freezer overnight and then pop off the brittle wax with your thumbnail or a small sharp knife.
- On carpeting: give it the frozen treatment with an ice cube and nick it off. If this doesn't work, place layers of absorbent paper and iron it off.
- On cloths: place folded recycled paper towels or blotting paper above and below the wax and press with a hot iron.
- On the floor: wash the floor with a mixture of ½ cup (125 mL) household ammonia and ½ cup (125 mL) detergent in a bucket of hot water. Wait until it's completely dry before rewaxing.
- On furniture: wait until the wax hardens or give it the ice cube treatment. Place a piece of aluminum foil over the wax and put a very hot towel on top. When it's hot enough, the wax should peel off easily.
- On the tablecloth: put the tablecloth in the freezer for an hour. Flick off all the wax you can, then place paper towels above and below the wax and iron with a relatively hot iron.

(*See also* Stain Removal: Fabrics.)

WEATHER STRIPPING

- If you suspect that there are drafts around windows, hold a lit candle up to them. If the windows aren't sealed properly, the candle will gutter. Use a hair dryer or heat gun to get the putty off. Reapply putty around the window: strip off the old putty and redo.
- Use plastic weather stripping as a temporary measure until you can do a more permanent job.

(*See also* Drafts.)

WEEDS

Weeds are nature's way of protecting precious soil. We go at them with chemicals and end up destroying the soil. Take a more benign point of view. If you have weeds, replace them with another plant. That and mulching are your two best bets in weed suppression.

- You can put a black cloth over a bed and cut slits for the plants, thus discouraging weeds from coming up. It's called landscape cloth. If you have a wooden frame, use nails to staple cloth down. Otherwise use rocks to keep the material in place.
- You can put mulch on top of the cloth, which will reduce weeds as well as reducing the amount of watering. (*See also* Mulching.)
- Try pour boiling water on crabgrass. Vinegar and salt poured directly on the weed are also effective.
- To keep grass away from spaces between brick, pour salt in those areas.
- Cut dandelions off close to the root and add mulch on top to kill off any growth.

WEEVILS

- You'll recognize them as tiny worms in food. Throw out anything contaminated and clean everything in the shelves that might be contaminated.
- Keep creepy crawlies away by putting bay leaves in flour, oatmeal and dry milk.
- Their favorite foods are beans and grains. Hang little containers, such as tea caddies of black pepper, around food-storage bins.

WET ROT (See Wood Rot.)

WINDOWS

Sealing Windows:

- Drive a nail in at an angle on either of the inside frames or get a special sash window lock, or install dual screws. This will keep sash from moving.
- Always seal windows from the inside against winter winds. Cover with plastic film from kits with enough material to cover 3 by 5 ft (1 by 1.5 m). Cut a piece 1 in (2.5 cm) larger than the window.
- Clean and dry the molding around window. The tape works best on aluminum vinyl-clad wood or painted wood. It doesn't work as well on plaster, wall board or veneered paneling.
- Make a frame with the tape on the flat spot of the molding.
- Use a hair dryer at its hottest setting to set the film. Hold ¼ in (6 mm) away from the film. The dryer creates shrinkage and smooths out wrinkles. Don't touch the film or you'll tear it. Trim excess.

Window Cleaning:

- Streaky windows are usually caused by a buildup of oily or soapy mixtures.
- Wash windows with a solution of one part white vinegar or lemon juice and five parts water. Spray solution on, then buff and clean with old newspapers.
- Window cleaner for very dirty windows: ½ cup (125 mL) household ammonia, 2 cups (500 mL) rubbing alcohol, 1 tsp (5 mL) phosphate-free liquid detergent. Add enough water to make 1 gallon (4 L).
- Never wash windows if the sun is shining directly on them as it causes streaking.

(*See also* Cleaning Formulas.)

WOOD PANELING

- To measure how many standard 4 by 8 ft (1.2 by 2.4 m) panels you will need, measure each wall's width and divide by 4 ft (1.2 m).
- Allow panels to adjust to a room's humidity and temperature for a couple of days before installing. It's preferable to leave them lying flat for this.
- The best quality is 4 by 8 ft (1.2 by 2.4 m) paneling with a thin veneer of real wood.

- The best panels are sequence-numbered or from the same skid so the same grain is repeated on each panel. When buying, stack panels side by side to see if they look good together. Make sure a groove matches the grain of the wood and isn't cut through the middle of it.
- To avoid warping, leave a little bit of space on the top and bottom and at the sides. Cover with molding.

WOOD ROT

Dry rot is a fungus that makes wood appear as though it's been stained white. Dampness provides the right condition for its spores to germinate. The fungus can spread and create serious damage.

- If you can, cut out any affected wood, get rid of it—burn if you have to—and replace. If the wood is in the walls, you may have a serious problem. This can happen when insulation is installed without a vapor barrier.
- Sterilize all wood or brickwork close to dry rot with a blow torch and treat with preservative once it's cool.

Wet rot, another fungus, shows up as dark stains on wood.

- Dry wood thoroughly, or cut out and replace.
- If you have a serious problem with either dry or wet rot, hire an expert. Get several quotes because there is a huge discrepancy in this area.

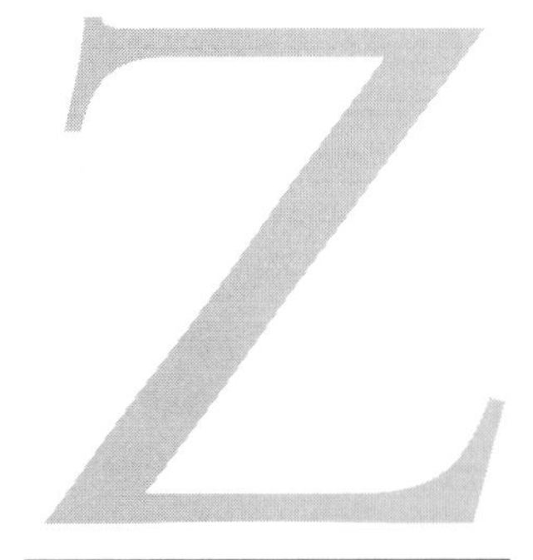

ZIPPERS

- Make a stuck zipper move by passing a bar of soap along its length, or rub candle stub on teeth for lubrication, or run a soft lead pencil up and down the teeth.

To Replace a Zipper:

- When you want to replace a zipper and can't find one of the exact same length, measure and mark the correct length and stitch over the coil to form a new bottom stop.
- If you need a bottom stop on a zipper that's too long, cut excess from the top and stitch both ends of tape.
- You can insert a zipper by hand—use tiny stitches. They may look fragile but are very strong. Run thread through candle stub to strengthen it and keep it from tangling.
- Before removing basting mark the line with transparent tape. Mark along the edge of the zipper tape for easy insertion of the replacement zipper.